BIOASSAY TECHNIQUES FOR DRUG DEVELOPMENT

BIOASSAY TECHNIQUES FOR DRUG DEVELOPMENT

Atta-ur-Rahman

and

M. Iqbal Choudhary
H.E.J. Research Institute of Chemistry
University of Karachi, Pakistan

and

William J. Thomsen
Arena Pharmaceuticals
San Diego, USA

harwood academic publishers
Australia • Canada • France • Germany • India • Japan • Luxembourg
Malaysia • The Netherlands • Russia • Singapore • Switzerland

 Printed in Singapore.

Amsteldijk 166
1st Floor
1079 LH Amsterdam
The Netherlands

British Library Cataloguing in Publication Data

A catalogue record for this book is available from the British Library.

ISBN 90-5823-051-1

CONTENTS

PREFACE

The so-called "green wave", triggered by a growing ecological awareness, has resulted in an increased interest in herbal formulations throughout the world, particularly in the last decade. The consumption of medicinal plants has almost doubled in the West during that period. The efficacy of a number of herbal formulations has been tested by valid phytopharmaceutical techniques and the number of plant-based drugs or health foods has increased steadily to meet the growing demand. Over the years a new relationship between phytochemists and pharmacologists has accordingly developed which, in many cases, has proved to be very productive.

Unfortunately, despite recent advances in chromatographic and spectroscopic techniques and the rich tradition of the use of herbal medicines, the majority of natural product chemists in developing countries are involved in empirical phytochemical practices and very little effort has been directed towards isolating the bioactive chemical constituents from the natural sources. This is due to the lack of expertise and infra-structure for biological screening and the often long waiting times required for such screening if samples are sent to other pharmacology laboratories. It is therefore highly desirable to establish in-house bioassays in phytochemistry laboratories which are inexpensive, rapid and do not require a specialized knowledge of biochemistry, biology or pharmacology. A number of phytochemical laboratories in the West have therefore established simple "bench-top" bioassays which can be carried out by non-specialists. The results obtained from such processes have strongly justified such a multidisciplinary approach. It is hoped that this manual of bioassay techniques will fulfill the need for a comprehensive text and prove useful to a large number of natural products chemists from around the world.

Bioassays can be divided into various broad groups based on the target life forms on which they are carried out. These could be:

1. Whole animals
2. Isolated organs of vertebrates
3. Lower organisms *e.g.* fungi, bacteria, insects, molluscs, lower plants, etc.
4. Cultured cells (such as cancer cells) and tissues of human or animal origin
5. Isolated subcellular systems, such as enzymes, receptors, etc.

The goal of an activity-directed isolation process is to isolate bioactive compounds which are capable of curing or alleviating a human or animal ailment and which can either be ultimately developed as established drugs directly or which can provide interesting structural leads. The process of drug development is long, tedious and expensive, requiring a multidisciplinary collaboration between botanists, pharmacognosists, chemists, pharmacologists and toxicologists, and clinicians. Simple and rapid bioassays can serve as starting points for such multidisciplinary efforts directed at drug discovery. The purpose of these bioassays is to rapidly screen for interesting biological activities, which can then be followed by more detailed mechanism-based studies of a multidisciplinary nature.

We therefore felt that there was a strong need of a book which would present the more important "bench top" bioassays. These can be integrated into the research programs by natural product chemists and thereby their efforts could be targeted to new drug discovery. A few specialized bioassays have also been included in the latter part of the book (e.g. anticancer screening using human cancer cell lines) for those phytochemistry laboratories having the desire, expertise and funds to implement these relatively more demanding techniques.

The second part of the book includes a number of enzyme-based assays, cell-based functional bioassays and receptor radioligand binding assays along with detailed descriptions of each type. The style of presentation of this section is deliberately more detailed as it constitutes rapidly developing technology and we feel that the material would be of considerable interest to the readers. However a major emphasis of the book is to present those bioassays which can be readily set up in any phytochemistry laboratory with limited funds, facilities or technical know-how. The majority of these bioassays have been presented in a step-wise format, the object being to make the procedures as simple as possible so that they can be implemented in chemistry laboratories by technical personnel with little background of microbiology, biochemistry or pharmacology.

We (A.R. and M.I.C) wish to acknowledge with thanks the secretarial help of Mr Mahmood Alam, Mr M. Asif and Mr S. Tauseef H. Naqvi. We are also grateful to Miss Zareen Amtul, Miss Zeba Parvin, Dr Mohsin Raza, Mr Usman Ghani, Mr Asaad Khalid, and Mr Atif Zaidi Lin (UCSD) who assisted us in the compilation of a number of bioassays and to Miss Shazia Anjum and Mr M. Yusuf who helped in proof-reading of the manuscript. One of us (M.I.C.) acknowledges the technical and financial assistance of Professor Jon Clardy (Cornell University, New York) during his stay in Ithaca, New York, in 1992 in which the major part of the literature cited in the book was collected.

1.0 GENERAL INTRODUCTION

Extracts from natural product sources have served as a valuable source of molecular diversity in many drug discovery programs, and several important drugs have been isolated from natural products. In any natural product isolation program in which the end-product is to be a drug or a lead compound, some type of bioassay screening or pharmacological evaluation must necessarily be used to guide the isolation process towards the pure bioactive component.

The pharmacological evaluation of extracts of organisms and pure isolates is an essential aspect of the drug discovery process and developments in the area of *in vitro* techniques have substantially transformed this facet of natural product chemistry. While it previously took weeks or even months to test a sample for some assays, it can now take only a few hours.

One should also distinguish between the "primary bioassay screens" from "secondary screens". Primary bioassays are assays which can be rapidly applied to a large number of samples to determine if any bioactivity of the desired type is present. They should therefore have high capacity, low cost and provide the results quickly. They need not be quantitative. "Secondary testing" procedures involve more detailed testing of lead compounds on a number of model systems in order to select compounds for clinical trials. They are usually low capacity, slow and costly assays.

The main requirements which a primary bioassay screen should meet are the following:

(1) The bioassay results should predict some type of therapeutic potential, either directly or by analogy with clinically effective drugs which have also been screened by the same procedure.

(2) Potentially useful pharmacological activity should not go undetected even though the activity may be either unexpected or unique.

(3) The probable nature of the activity should be indicated so that subsequent research can be organized intelligently.

(4) The primary bioassay screening test should be tolerant of the many impurities present in a crude extract and yet it should be sensitive enough to reveal presence the potentially interesting substances present in low concentrations (levels of about

0.0001% of an active compound in an extract, based on the dried weight of the extracted organism, should be detectable).

(5) The bioassay procedure should be unbiased and it should allow for the coding of all samples, including both "known" reference materials (standards) and "unknown" test samples.

(6) The results obtained should be reproducible.

(7) The screen should allow the use of both crude materials and pure isolates so that the procedure can be used to direct the extraction, isolation and purification work of the natural product chemist.

(8) Completion of a single bioassay screen should not require more than 1.0-2.0 g of the crude dry natural material (plant or animal extract).

(9) The primary bioassay screen should have a high throughput, even if the information content is low, with the results becoming available quickly.

(10) The procedure should not require expensive equipment or a sophisticated laboratory environment so that the primary level screening experiment can be conducted synchronously with the fractionation process.

(11) The procedure should be compatible with the use of dimethyl sulfoxide (DMSO) since DMSO is commonly employed to solubilize extracts or pure polar compounds for screening.

(12) The procedure should be simple enough to be taught easily to laboratory technicians so that highly trained and qualified researchers are not required for the routine operation of the bioassay program.

(13) The test animals (if required for the bioassay) should be easily obtainable, easily handled, easily bred and resistant to infections.

(13) Finally the bioassay should be economical to conduct over extended time periods.

A hit rate of 1% or less is generally considered a reasonable and one then proceeds from primary screening to secondary screening for profile selectivity and in order to establish biological activity.

The bioassay-guided natural product drug discovery research can be broadly divided into four approaches:

(1) the use of a single bioassay technique in order to search for a specific type of pharmacological activity (such as antidiabetic, cardiotonic or antiinflammatory activity);

(2) the use of a battery of specific bioassay techniques with each procedure directed to discover a different type of useful activity;

(3) the use of a single bioassay technique designed to detect multiple activities (non-specific bioassay). For example cytotoxicity bioassays can be employed to predict a variety of biological activities such as antitumor, insecticidal and antimicrobial activities. Another example of this approach is the study of drug-induced symptomatology (CNS depressant, tranquilizing, psychotropic, skeletal muscle relaxant, sympathetic stimulant, diuretic, metabolic poison, vasodilatory etc.) after a single intraperitoneal injection of an extract to an intact anaesthetized rat;

(4) the use of a combination of a variety of bioassays in order to detect specific activities as well as to detect multiple activities.

The choice of the screening approach to be adopted generally depends on the target disease as well as on the available information about the target organism (plant, marine animal, etc.) to be studied. For example if a plant has a ethanopharmacological history of use against a particular disease, then one would logically use a specific bioassay technique (single goal screening) which can predict the reputed therapeutic activity in order to isolate the lead which is responsible for that biological activity. Similarly, based on the chemotaxonomic knowledge of related species, one can select one or more bioassay screens based on the reputed or reported bioactivity (or use) of related species. Bio-rational selection is based on the knowledge of plants and animals and their behavior in certain circumstances. For example some primates may eat certain types of grasses in cases of indigestion. The zoo-pharmacognosic knowledge can hence help in the selection of specific bioassays. Similarly certain plants many exhibit resistance against insect attack. They can therefore be screened for insecticidal compounds by

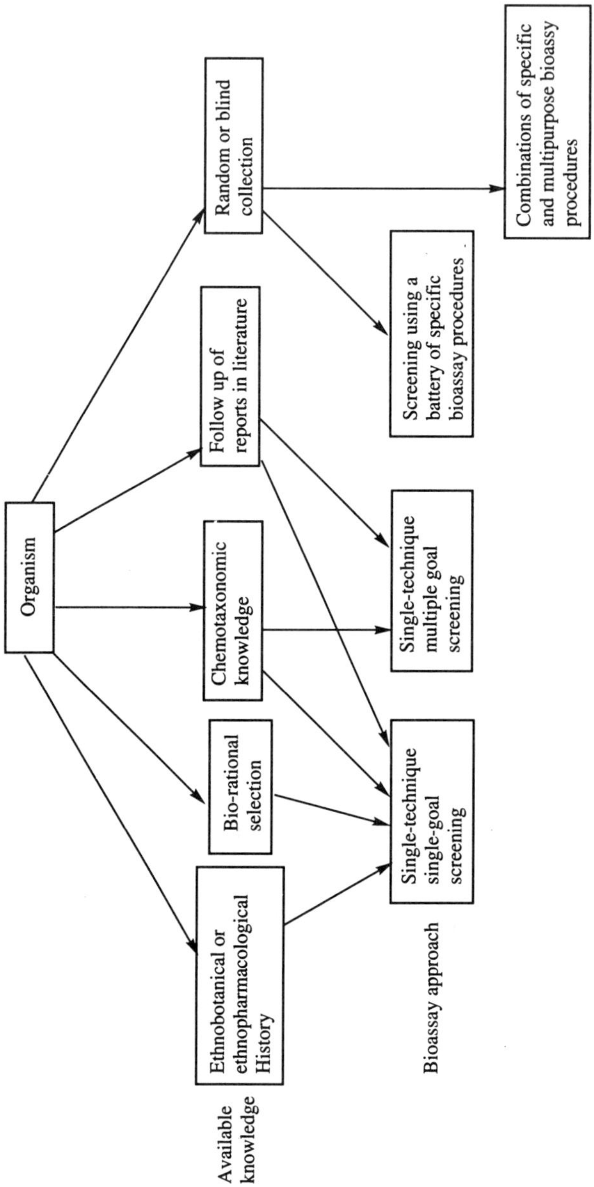

Scheme-1: Approaches to bioactivity-directed natural product drug development.

using pesticidal bioassays. In the case of random or blind collection of common, unusual or uninvestigated organisms, it is better to use a battery of bioassay screens at the extract level and follow the most prominent activity subsequently through a specific bioassay technique (Scheme-**1**). A typical flow diagram of a bioassay-guided isolation of bioactive isolates from natural sources (plants, microorganisms, marine animals, etc.) is presented in (Scheme-**2**).

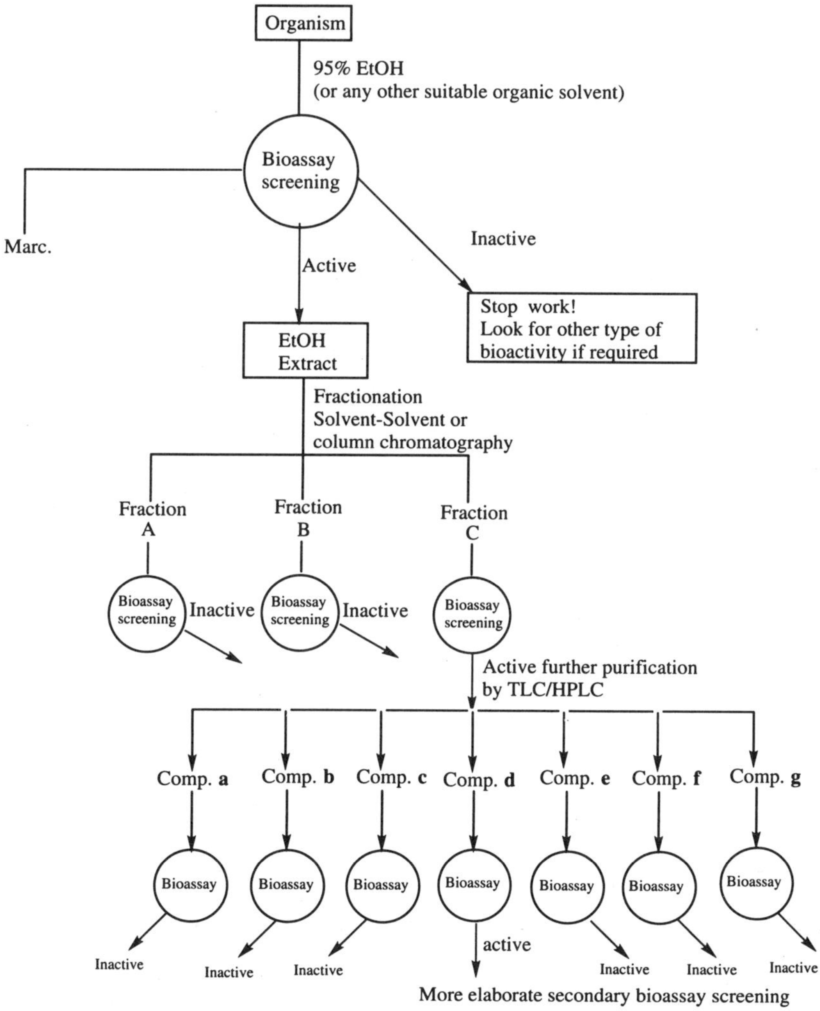

Scheme-2: Bioactivity-directed isolation of natural products.

PART-A

Bench-Top and Primary Bioassay Screening

1.1 TOXICITY ASSAYS

1.1.1 Brine-Shrimp Lethality Assay

Bioactive compounds are often toxic to shrimp larvae. Hence, *in vivo* lethality to shrimp larvae can be used as a rapid and simple preliminary monitor for bioactive compounds during the isolation of natural products. The eggs of the brine shrimp *Artemia salina* (Leach) are readily available as fish food in pet shops. When placed in artificial sea water, the eggs hatch within 48 hours, providing large numbers of larvae. These tiny shrimp larvae have been extensively used as a tool to monitor the cytotoxicity of samples under study. This is a rapid, inexpensive, in-house, general bioassay which has been developed for screening, fractionation and monitoring of physiologically active natural products (Meyer *et al.*, 1982).

Materials

1. *Artemia salina* Leach (brine shrimp eggs)*
2. Sea salt+
3. Small tank with perforated dividing dam and cover to grow shrimps; lamp to attract shrimps
4. Syringes; 5.0 ml, 0.5 ml, 100 µl and 10 µl
5. 2 dram vials (9 per sample + 1 control)
6. Magnifying glass
7. Organic solvents (methanol, dichloromethane, chloroform, DMSO etc.)
8. Distd. water
9. Pasteur pipettes
10. Aluminium foil
11. Test sample (crude extract of organism, pure natural product or synthetic compound)

The steps involved in the Brine-shrimp lethality assay are as follows:

1. Artificial "sea water" is prepared by dissolving *ca.* 3.8 g sea salt per liter of water and filtered.

---

* Available from San Francisco Bay Brand, Inc., New York, CA 94560, U.S.A.

\+ Available from Instant Ocean, Inc., 8141 Tyler Boulevard, Mentor, OH 44060, U.S.A.

2. "Sea water" is placed in a small unequally divided tank and shrimp eggs added to the larger compartment of the tank which is darkened by covering it with aluminium foil. The illuminated compartment attracts shrimp larvae (nauplii) through perforations in the dam.
3. Allow 2 days at room temperature (22-29 °C) for the shrimps to hatch and mature.
4. Prepare vials for testing; for each fraction, test initially at 1000, 100, and 10 μg/ml; prepare 3 replicates for each concentration making a total of 9 vials; weigh 20 mg of sample and add 2 ml of organic solvent (20 mg/2 ml); from this solution transfer 500, 50, or 5 μl to vials corresponding to 1000, 100, or 10 μg/ml, respectively. Evaporate solvent under nitrogen and then place under high vacuum for about 30 min; the volatile organic solvents will evaporate overnight. Alternatively, polar insoluble materials may be dissolved in DMSO, and upto 50 μl may be added per 5 ml of "sea water" before DMSO toxicity affects the results.
5. After 2 days (when the brine shrimp larvae have matured), add 5 ml "sea water" to each vial and add 10 shrimps per vial with the help of Pasteur pipette (30 shrimps per dilution). The vials are maintained under illumination.
6. After 24 hours have elapsed, count and record the number of surviving shrimps, with the aid of a 3x magnifying glass.
7. Analyze data with a Finney computer program (Probit analysis) to determine LC_{50} values and 95% confidence intervals.**
8. Additional dilutions of less than 10 μg/ml may be needed for potent materials. Intermediate concentrations can be prepared and tested to narrow the confidence intervals.

1.1.2 Brine-Shrimp Microwell Cytotoxicity Assay

A new microplate assay for cytotoxicity or lethality determination using brine-shrimp (*Artemia salina*) has been developed which gives results comparable to the vial method described under the heading of brine-shrimp lethality assay (Section 1.1.1). The assay reliably correlates with KB cell toxicity assays and thus provides a convenient means by which

---

** A copy of this program for IBM PC's is available on request from Prof. Jerry L. McLaughlin, Department of Medicinal Chemistry and Pharmacognosy, Purdue University, West Lafayette, IN 47907, U. S. A.

the presence of cytotoxic natural products may be detected during the fractionation and isolation of natural products.

Materials

1. Brine shrimp eggs (*Artemia salina*)
2. Sea salt
3. Dried yeast
4. 96-Well microplates
5. DMSO
6. Pasteur pipette
7. Binocular microscope (10.30x)
8. Methanol
9. Incubator
10. Beaker
11. Test sample (plant extract, pure natural product or synthetic compound)

Brine-shrimp microwill cytotoxicity assay typically consists of the following assay steps:

1. Artificial sea water is prepared by dissolving sea salt in distd. water (40 g/lit.) supplemented with 6 mg/lit. dried yeast.
2. Brine shrimp eggs (*Artemia salina*) are hatched in artificial sea water during 48 hours incubation in a warm room (22-29 °C).
3. Brine shrimp larvae (nauplii) are collected with a Pasteur pipette after attracting the organisms to one side of the vessel with a light source. Nauplii are separated from the eggs by pipetting them 2-3 times in small beakers containing sea water.
4. The test sample (20 mg of crude extracts or 4 mg for pure compound) is made up to 1 mg/ml in artificial sea water (water insoluble compounds or extracts can be dissolved in 5 ml DMSO prior to adding sea water).
5. Serial dilutions are made in wells of 96-well microplates in triplicate in 100 μl sea water.
7. A suspension of nauplii containing 10-15 brine shrimp larvae (100 ml) are added to each well with the help of a Pasteur pipette and the covered microwell plate incubated at 22-29 °C for 24 hours.
8. The plates are then examined under a binocular microscope (12.5x) and the number of dead (non-mobile) nauplii in each well counted.
9. 100 μl methanol is then added to each well and after 15 minutes the total number of shrimps in each well is counted.

10. LC_{50} values are then calculated by using Probit analysis ("FIN" program).

1.1.3 Crown Gall Tumor Inhibition Assay (Potato Disc Antitumor Assay)

Crown gall is a neoplastic disease of plants which is induced by the gram negative bacteria *Agrobacterium tumefaciens*. The bacteria possess large Ti (tumor inducing) plasmids which carry genetic information (T DNA) that transform normal, wounded, plant cells into autonomous tumor cells. Since the mechanism of tumor induction is similar to that in animals, this test system has been used to evaluate and pre-screen the antitumor/cytotoxic properties of natural products. The results suggest that the potato disc assay is a safe, simple, rapid and inexpensive in-house screen for 3PS antitumor activity. It is statistically more predictive of 3PS (P 388 leukemia) activity than either the 9KB (human nasopharyngeal carcinoma) or 9PS (murine leukemia) cytotoxicity assays. The assay also gives indication of tumor-promoting or carcinogenic properties of the test samples (Ferrigni *et al.*, 1982).

Materials

1. Laminar flow hood or clean air chamber
2. Fresh, disease-free potato tubers (preferably red).
3. Organic solvents (DMSO and ethanol)
4. Broth culture of *Agrobacterium tumefaciens* strain B6 (ATCC)* .
5. 1.5 Sterile cork borer
6. Millipore filters (0.22 mm)
8. Disposable or autoclavable gloves
9. Difco agar
10. Autoclave
11. Sterile petri dishes
12. Pipettes
13. Test tubes
14. Large glass tray
15 Aluminium foil
16. Parafilm
17. Liquid hypochlorite bleach
18. Sterile special cutter (3 cm long small knives fixed parallel with each other on a wooden frame with holder) or scalpel
19. Incubator
20. Dissecting compound microscope

---

* Available from American Type Culture Collection. Rockville, Maryland, U.S.A.

21. Lugol's solution (5% I_2 + 10% KI in H_2O)
22. Test sample (crude extract, pure natural product or synthetic compounds)

The potato disc antitumor assay involves the following steps:

1. Fresh potato tubers (disease free) of moderate size are sterilized by soaking in liquid bleach for 10 minutes.
2. A core cylinder of tissue is removed from the potato by means of a sterilized cork borer. Two cm ends of each potato cylinder should be discarded and the remainder of the cylinder is cut into discs of uniform thickness (0.5 cm) by a special cutter or scalpel under aseptic conditions.
3. The potato discs are then transferred to 1.5% agar plates (1.5 g of agar/100 ml distd. water, autoclaved and 20 ml of agar solution poured in each sterile Petri dish). Five potato discs should be placed in each Petri dish and 3-5 dishes are used for each test sample along with the same number of dishes for the control.
4. 8 mg of sample is dissolved in 2 ml of DMSO in a test tube and filtered through a Millipore filter into another sterile tube. 0.5 ml of this solution is then added to 1.5 ml of sterile autoclaved distd. water, and then 2 ml of a broth culture of *A. tumefaciens* (a 48 hour culture containing $5x10^9$ cell/ml) is added.
5. Controls are prepared by filtering 0.5 ml DMSO through a Millipore filter into 1.5 ml of sterile distilled water and adding to tubes containing 2 ml of a broth culture of *A. tumefaciens.*
6. 1 drop (0.05 ml) is drawn from these test tubes using a sterile pipette, and it is used to inoculate each potato disc, spreading it over the disc surface. The process starting from the cutting of the potatoes to the inoculations should be completed within 30 min in order to avoid contamination[+] .
7. The Petri dishes are incubated at room temperature (27 ^oC), the lids being taped down by using parafilm to minimize moisture loss.
8. After twelve days of inoculation, the tumors are counted with the aid of a dissecting microscope after staining with Lugol's solution (the tumors can also be counted without using Lugol's solution). The tumor cells lack starch. The number of tumors in the control are used as a reference for activity.
9. The results are derived from the number of tumors on test discs versus those on the control discs. Inhibition is expressed as a

---

[+] It has been observed that if the procedure takes a longer time, the workers may lose concentration and the chances of contamination can increase.

negative percentage and stimulation is expressed as a positive percentage. 20% inhibition in two or more independent assays is considered as significant activity of a test sample.

1.1.4 Animal Toxicity Assay

Materials

1. BALB/c mice (30 mice per test sample and 6 mice for control)
2. Syringes
3. Saline solution (0.85% sterile NaCl)
4. Autoclave
5. Test sample (plant extract, pure natural product or synthetic compound)

The following steps are involved in animal toxicity assay:

1. Six groups of five mice each are injected intraperitonially with different dilutions of test sample (50, 100, 150, 200 and 250 mg dissolved in saline).
2. The control group of the animals is only administered sterile saline.
3. The animals are kept in observation for one week and deaths of animals are recorded.
4. LD_{50} is calculated by the standard method (Kazmi *et al.*, 1990).

1.2 ANTIMICROBIAL ASSAYS

A large number of human, animal and plant disease are caused by pathogenic microbes (fungi bacteria and algae). Infection due to fungi and bacteria have been a major cause of death in higher organisms. The discovery of antibiotic penecillin by Fleming is therefore considered to be one of most important discoveries in the world. Historically many of the new antibiotics were isolated from natural sources (soil microbes, plants, etc). Many more were later synthesized and introduced in clinical practices. Unfortunately human struggle against pathogenic microbes is far from over due to many reasons. Most important of them time to time discovery of new pathogens, and remarkable abilities of microbes to develop resistance against used antibiotic. The discovery and development of new antimicrobial agent is therefore a on going process. Remarkable diversity of chemicals present in biological sampleshave tremendous potential in search of new antimicrobial agents.

In order to prepare plant extracts for antimicrobial testing, (*i.e.* testing for antibacterial and antifungal activity) the fresh plant material (or dried powdered material) may be macerated or percolated with water or an organic solvent. The antimicrobial testing may be carried out either on these crude extracts or after separation into further sub-fractions or pure compounds. A bioassay-directed protocol is recommended in which the activity is followed from the crude extract through the various fractions, leading finally to the pure active compounds. Care must be taken to avoid heat during the evaporation of the solvent from the extract or from various fractions in order to prevent thermo-labile antimicrobial agents present from being destroyed. Extracts should therefore not be subjected to autoclaving (for sterilization). Sterilization by membrane filtration should also be avoided since antimicrobial compounds can also stick to the membrane surface, thereby rendering the extract inactive. Solvents being used should be tested as controls in order to ascertain that the antimicrobial activity is not due to the solvent. The pH of the extract or fractions should be checked since microorganisms may not be able to grow in media which are excessively acidic or basic. It is advisable to make the extracts neutral (pH 6.0 to pH 8.0) prior to testing. Alternatively they can be dissolved in a physiological Tris buffer or some other buffer solution. Sterilization of the samples prior to testing can be carried out by gamma irradiation but this facility may not be present in many laboratories and the method is time consuming. Generally it is best to extract the plant material using 80% EtOH-20% H_2O as this will kill most bacteria present and usually serve to extract both organic solvent-soluble and water-soluble compounds. The ethanol must be removed completely before testing for antimicrobial activity.

There is no single all-embracing bioassay to evaluate the antimicrobial activity of a sample. Therefore, the evaluation process generally involves the use of a number of bioassay methods and careful comparison of all the data in order to arrive at an appropriate conclusion (Linton, 1983).

There are three major methods for antimicrobial testing: (a) agar diffusion method (b) agar dilution method (c) bioautographic method.

In the *agar diffusion method*, wells are cut in seeded agar and the test sample is then introduced directly into these wells. After incubation the diameter of the clear zone around the well is measured and compared against zones of inhibition produced by solutions of known concentrations of standard antibiotics. In samples where the presence of suspended particle matter (or precipitation of water-insoluble substances on the disc or cylinder) interferes with the diffusion of the antimicrobial substance, warming on a hole-plate may be advantageous. Five or six samples may be tested simultaneously by the diffusion method.

In the *agar dilution method*, the medium is inoculated with the test organism and the samples to be tested are mixed with the inoculated medium. The material is inoculated and the growth of the microorganisms is viewed and compared with a control culture which does not contain the test sample. The experiment is repeated at various dilutions of the test sample in the culture medium and the highest dilution at which the sample just prevents the growth of the microorganism (MIC) is determined.

The *bioautographic procedure* for screening for antimicrobial activity involves localizing the antibacterial activity on a chromatogram. The antimicrobial agent is transferred from the TLC plate or paper chromatogram to an inoculated agar plate by diffusion and the zones of inhibition visualized.

It is important to mention here that all manipulation of microbial material should be performed in a contained environment (Laminar flow chamber or glove box) with disposible surgical gloves. This would minimize the chances of any infection to worker or contamination in test procedure.

It is also important that all contaminated materials should be collected in antoclave bags and autoclaved at 120 oC before disposal. Laboratories involved in antimicrobial screenings are also encouraged to develop their own biosafety rules based on standard procedures.

1.2.1 Antibacterial Assays

1.2.1.1 Agar diffusion assay

Materials

1. Test organisms, e.g. *Escherichia coli* (NCTC 10418), *Bacillus subtilis* (NCTC 8236), *Staphylococcus aureus* (NCTC 6571), *Pseudomonas aeruginosa* (ATCC 10145) (**Caution!**)*.
2. Nutrient broth
3. Sterile cork borers
4. Petri dishes (14 cm)
5. Pipettes (0.1 ml and 1 ml)
6. Organic solvent
7. Incubator
8. Standard antibiotics (streptomycin, ampicillin, etc.)
9. Test sample (crude extract, pure natural product or synthetic compound)

The agar diffusion assay consist of the following steps:

1. 10 ml aliquots of nutrient broth is inoculated with the test organisms and incubated at 37 °C for 24 hr.
2. Using a sterile pipette, 0.6 ml of the broth culture of the test organism is added to 60 ml of molten agar which has been cooled to 45 °C, mixed well and poured into a sterile Petri dish (for the 9 cm Petri dish, 0.2 ml of the culture is added to 20 ml of agar). Duplicate plates of each organism is prepared.
3. The agar is allowed to set and harden and the required number of holes are cut using a sterile cork borer ensuring proper distribution of holes (cups) in the periphery and one in the center. Agar plugs are removed. Different cork borers should be used for different test organisms.
4. Using a 0.1 ml pipette, 100 μl of the test sample dissolved in an appropriate solvent is poured into appropriately labelled cups (these are marked at the back of the cup before filling). The same concentrations of the standard antimicrobial agents (streptomycin 1 mg/ml and ampicillin 10 μg/ml) and the solvent (as control) are used.
5. The plates are left at room temperature for 2 hr. to allow diffusion of the sample and incubated face upwards at 37 °C for 24 hr.
6. The diameter of the zones of inhibition is measured to the nearest mm (the cup size also being noted) (Kavanagh, 1963; Leven *et al.*, 1979).

1.2.1.2 Agar dilution assay

Materials

1. Test bacteria such as *Bacillus subtilis, Staphylococcus aureus, Streptococcus faecalis, Escherichia coli, Agrobacterium tumefaciens, Klebsiella pneumoniae, Pseudomonas aeruginosa, Proteus vulgaris,* etc. (**Caution** !).*
2. Nutrient agar (composition g/l, peptone 5 g/l, NaCl 5 g/l, beef extract 15 g/l, yeast extract 15 gm/l, pH 7.2, agar 20 gm/lit).
3. Organic solvents (ethanol or acetone)
4. Test tubes
5. Incubator
6. Pipettes (0.5 ml, 1 ml, 10 ml).
7. Test sample (crude extract, pure natural product or synthetic compound)

The following steps are involved in agar dilution assay:

1. A loopful of the bacterial culture[#] from the slant is inoculated in the nutrient broth and incubated at 37° ± 1 °C for 24 hours.
2. The fresh broth (20 ml) is seeded with 0.25 ml of the 24 hour broth cultures and a two-fold serial dilution method is followed as described below. The test sample is dissolved in water (in case of water-soluble samples) or in an organic solvent (ethanol or acetone) to obtain a 10 mg/ml solution. A 0.2 ml solution of the test material is added to 1.8 ml of the seeded broth and this forms the first dilution.
3. 1 ml of this dilution is diluted further with 1 ml of the seeded broth to produce the second dilution, and the process is repeated until six such dilutions are obtained.
4. A set of tubes containing only seeded broth is kept as control and suitable solvent controls are also maintained.
5. After incubation for 24 hr at 37° ±1 °C the last tube with no visible growth of the microorganism is taken to represent the minimum inhibitory concentration (MIC) of the test sample which is expressed in mg/ml.

1.2.1.2.1 Microtitre plate method:

If a 96-well microtitre plate is available then three dilutions of a test sample can be tested against 24 microorganisms (or two extracts against 12 microorganisms) simultaneously. In that case the following procedure may be followed:-

1. The test sample is dissolved/suspended in a physiological Tris buffer (pH 7.2) or in a mixture of polyethylene glycol 400 (PEG 400) and physiological Tris buffer (4:6) (15 ml).
2. The solubilized test sample (2.0 ml) (or its four-fold dilutions, 1/4, 1/16, etc.) is warmed and mixed with an equal amount of liquid agar medium at 50 °C, thereby affording the dilutions 1/2, 1/8, 1/32 , etc.
3. The microtitre plate is warmed with an infra-red lamp and the holes of lanes B-D or F-H are filled with these dilutions (0.3 ml per hole). The holes in lanes A and E are filled with the control consisting of a mixture of the solubilizing buffer (2.0 ml) and culture medium (0.2 ml to 0.3 ml per hole).

[#] Now many bacteria can also be purchased as dried discs from several companies. Some bacteria are highly pathogenic. Complete details for biosafety precautions should be obtained from the supplier of microbes or from qualified microbiologists.

4. The infra-red lamp is removed to allow the materials to solidify at room temperature and all the holes are then inoculated with 1:100 dilution (5 μl) of overnight cultures of test bacteria (± 10^3 bacteria).*
5. Inoculation is carried out for 24 hour at 36 °C, and a light microscope is then used to compare the growth of test organisms against the control. Test samples showing inhibitory effects at all three dilutions (1/2, 1/8 and 1/32) are subjected to further investigations (Srivastava, 1984).

Caution! Some of these microbes are human pathogens.

1.2.1.3 Direct Bioautography Method

Bioautography can be employed as a method for localizing antibacterial activity on a chromatogram. The "agar diffusion" technique involves transfer of the antibacterial compound from the chromotographic plate to an inoculated agar plate by diffusion, and visualisation of the zones of inhibition. An improved simpler version of this method, which avoids extensive microbiological equipment and problems associated with differental diffusion of compounds from the chromatogram to the agar plate, is presented below (Hamburger, 1987). In this method a suspension of a microorganism in a suitable broth is applied to tlc plate which is then incubated in a humid atmosphere to allow growth of the bacteria. Zones of inhibition are then visualised by a dehydrogenase-activity detecting reagent, a tetrazolium salt, which is converted by the bacteria into the intensely colored product. The antibacterial compounds appear as colorless spots against a colored background.

A. *Preparation of bacterial suspension*

Materials
1. Culture flasks (1 liter)
2. Inoculation loops
3. Shaker bath
4. Laminar flow hood
5. Disposable centrifugation tubes (50 ml)
6. Centrifuge
7. Colorimeter
8. Disposable pipettes (5, 10 ml)

---

* Available from American Type Culture Collection (ATCC).

9. Nutrient broth and nutrient agar (BB1)
10. Culture tubes
11. ATCC cultures of *B. subtilis* (ATCC # 6633) and *E. coli* (ATCC # 25922)† (**Caution !**)
12. Test sample (crude extract, pure natural product or synthetic compound)

The direct bioautography method consists of following steps:
1. Lyophilized ATCC bacterial culture is resuspended in the recommended broth (see instruction sheet of ATCC delivered with the culture). Agar slants are then inoculated. The bacteria can be kept on agar slants at 4 °C, and should be checked for viability and contamination every month. For safety reasons it is recommended to store a part of the initial inoculum in liquid nitrogen as a backup in case the slants are accidentally contaminated.
2. Nutrient broth (NB) (300 ml in 1 liter flasks) is inoculated with *B. subtilis* or *E. coli* (maintained on agar slants) and kept at 37 °C on a shaker at 80 rpm for 36 to 48 h.
3. The suspension is centrifuged for 10 min at 1500 rpm (in 50 ml centrifugation tubes) and the supernatant is discarded.
4. The bacteria is resuspended in 3-4 ml of fresh NB (S1). 2 ml of suspension is diluted to 20 ml by adding NB; the turbidity is determined by measuring the absorbance at 560 nm.
5. The solution S1 is diluted such that turbidity of a 1:10 dilution has approx. 0.84A (approx. 10^9 bacteria/ml).
6. The material is dispensed in cryovials (2 ml) and stored in liquid N_2.
7. Aliquots of 18 ml NB are dispensed in sterile culture tubes. These tubes are then stored at 4 °C for some weeks.

B. *Bioautography*

Materials
1. Aluminium backed silica gel TLC sheets, GF 254 (Merck)
2. Disposable surgical gloves
3. Autoclavable polyethylene boxes (larger than 20 x 20 cm)
4. Chromatography paper (Whatman)
5. Pyrex glass dishes (rectangular, 10 x 15 cm)
6. Autoclave bags (Fisher Scientific)
7. TLC sprayer, glass
8. TLC TANK
9. Ethanol

11. Lysol disinfectant
12. Roller device, 10 cm (should be autoclavable)
13. Double-sided adhesive tape
14. Safety hood, glove box or glove bags (Aldrich Chemicals)
15. Polyethylene stoppers (autoclavable)
16. Syringe with luer lock (20 ml)
17. Disposable membrane filters with luer lock
18. Disposable graduated TLC micropipettes (glass, 5 μl)
19. *p*-Iodonitrotetrazolium violet (INT) (Sigma)
20. HPLC grade water (fresh from Millipore unit)
21. Cryovial (2 ml) of conc. bacterial suspension
22. Nutrient broth (18 ml)
23. Test sample (crude extract, pure natural product or synthetic compound)

The steps involved in bioautography assay are as follows:

1. TLC chromatograms* (20 x 20 cm) are developed in a suitable solvent system (the mobile phase has to be sufficientlys volatile so that it can be removed completely.Traces of organic solvents will otherwise inhibit bacterial growth).
   Suitable organic solvents include: $CHCl_3$, CH_2Cl_2, MeOH, H_2O, CCl_4, i-PrOH, acetone, ether (avoid: toluene, benzene, *n*-BuOH, acids and bases).
2. TLC are run in duplicate (one for the bioautogram, one for comparison)
3. Both plates are dried carefully with a hair-drier and UV absorbing spots are marked on both plates (UV 254 + 366 nm). The silica layer of the plate should not be touched, specially the one which will be used for bioautography (to avoid contamination of the plate!). A control TLC plate is also kept. The silica layer of the control TLC plate should be covered in order to minimize the oxidation of compounds. The control TLC can be sprayed with a suitable chromogenic reagent such as phosphomolybdic acid, ceric sulfate, vanilline, Dragendorff's reagent or simply placed in a tank with iodine crystals the following day. The stained control TLC should be covered with parafilm to avoid discaloration.
4. A roller device is wrapped with a layer of chromatography paper of suitable size and fixed with double sided adhesive tape.

*Substantial quantities of crude extract (or fractions) should be spotted on the TLC plates, since if the test sample is too little, the chances of missing the active constituents will increase. On the other hand, a very high concentration of the test sample may lead to imperfect separation on TLC. The recommended quantity of crude extract is 10-100 mg/spot.

5. Polyethylene box (autoclaved) is lined with the chromatography paper, and soaked with fresh HPLC grade water (from Millipore unit) (approx. 20-30 ml). Polyethylene stoppers (6-8) and placed in the box (the stoppers will support the TLC plate and keep it separate from the wet paper).
6. The cryovial (2 ml) is thawed quickly at 37° C; 18 ml NB is added and poured into Pyrex glass dish.
7. The roller device is carefully soaked with the bacterial suspension and gently applied onto the silica layer. The process is repeated until the silica is soaked with the liquid (approx. 5-6 ml for TLC plate 20 x 20 cm). (Another approach: dip the plate into the pyrex dish and let the excess of liquid pour off. The zones of inhibition may however be less sharply visible by this method).
8. The bioautogram is placed into the polyethylene box and the lid is closed. The exterior is rinsed with EtOH 70%.
9. Incubate overnight at 37 °C.
10. Aqueous solution of INT (20 mg/ml) is prepared and filtered through the membrane filter into the glass TLC sprayer.
11. The bioautogram is sprayed with approx. 5 ml of TTC solution. The plate can be sprayed directly in the polyethylene box.
12. Incubate for 4 h at 37 °C.
13. Open lid, spray bioautogram with approx. 5-10 ml EtOH, 70%.
14. Stain control TLC with suitable chromogenic reagent.
15. The bioautogram is evaluted by comparison with the control TLC. Zones of inhibition (indicating the presence of antibacterial compound) on the bioautogram appear as white spots on a pink background.

Documentation by Polaroid photography is recommended (optional)

CAUTION!: All manipulations of bacterial materials should be performed in a contained environment (Laminar flow chamber or glove box) with disposable surgical gloves. All contaminated materials should be collected in autoclave bags and autoclaved at 120 °C before disposal.

1.2.2 Antifungal Assays

1.2.2.1 Agar tube dilution assay

There is considerable need to discover new fungitoxic compounds in view of the many plant and human fungal diseases. Some of the common plant fungal diseases are potato late blight, tobacco blue mould, hop downy mildew, Dutch elm disease, ergot of rye, cereal rusts, corn blight and grape downy mildew. The human fungal diseases include athletes foot, aspergillosis, actinomycosis, histoplasmosis and corcidiomycosis. Some fungi can be beneficial to man since they attack harmful insects (Blank *et al.*, 1965; Brass *et al.*, 1979).

Materials

1. Test fungi (mostly dermatophytes) such as *Epidermophyton floccosum, Trichophyton mentogrophytes, T. rubrum, T. simii, T. schoenleinii, Microsporum canis, Pseudallescheria boydii, Candida albicans,* etc. (**Caution!**)*
2. Sabouraud dextrose agar (composition in gm/l, pepto complex 10, glucose 40, agar 15).
3. Dimethyl sulfoxide (DMSO)
4. Screw test tubes
5. Incubator
6. Micropipettes
7. Magnetic stirrer
8. Autoclave
9. Standard antifungal drugs such as amphotericin-B, miconazole, ketoconozole, flucytopsine etc.
10. Test sample (crude extract, pure natural product or synthetic compound)

The following steps are involved in agar tube dilution assay:

1. Test sample is dissolved in sterile DMSO to serve as stock solution.
2. Sabouraud dextrose agar is prepared by mixing Sabouraud 4% glucose agar and agar agar in distilled water.
3. It is then stirred with a magnetic stirrer to dissolve it and a known amount is dispensed into screw capped test tubes.
4. Test tubes containing media are autoclaved at 121 °C for 15 minutes.
5. Tubes are allowed to cool to 50 °C and the test sample of desired concentrations pipetted from the stock solution into the non-solidified Sabouraud agar media.
6. Tubes are then allowed to solidify in a slanting position at room temperature.
7. Each tube is inoculated with a 4 mm diameter piece of inoculum removed from a seven day old culture of fungi (1).

***Caution !** Pathogenic

8. All culture containing tubes are inoculated at optimum temperature of 28-30 °C for growth for 7-10 days. Humidity (40% to 50%) is controlled by placing an open pan of water in the incubator.
9. Cultures are examined atleast twice weekly during the incubation.
10. After the incubation for 7-10 days, the test tubes with no visible growth of the microorganism is taken to represent the minimum inhibitory concentration (MIC) of the test sample which is expressed in µg/ml.

1.2.2.2 Direct bioautobiography method

This is an elegant method was designed for the detection of fungitoxic substances. This method is particularly suitable for rapid separation of antifungal substances from natural sources (Homans *et al.*, 1970).

Materials

1. TLC plates (silica gel)
2. Inorganic stock solution. The following inorganic salts are dissolved in 1 liter of dist. H_2O:-

KH_2PO_4	7 g
$Na_2HPO_4.2H_2O$	3 g
KNO_3	1 g
$MgSO_4.7H_2O$	1 g
NaCl	1 g

3. UV lamp
4. Test fungi such as *Aspergillus niger, Ascochyta pisi, Botrytis cinerea, Colletrichum lindemuthianun, Fusarium culmorum, Penicillium expansum, Glomerella cingulata, Cladosporium cucumerinum*[+] (**Caution !**)*.
5. Organic solvents
6. TLC tanks
7. Autoclave
8. Glucose
9. Incubator
10. Hot air blower
11. Test sample (crude extract, pure natural product or synthetic compound)

***Caution !** Pathogenic.

[+] Available from American Type Culture Collection (ATCC), Rockville, Maryland, U.S.A.

Here are the steps which involved in direct bioautography method:

1. The TLC chromatogram is developed in a TLC tank by using a suitable solvent system. Volatile organic solvents are generally used and the chromatograms are dried by using a hot air blower.
2. UV absorbing spots are located and marked under a UV lamp.
3. After locating the UV absorbing spots, the chromatograms are usually sprayed with a conidial suspension (caution). During the spraying, care should be taken to avoid the plates becoming too wet.
4. The conidial suspension containing *Cladosporium cucumerinun* (or any other fungus against which activity is needed to be determined) is taken up in the inorganic medium. The solution is autoclaved at 120°C for 20 min. Just before making the conidial suspension, 10 ml of a 30% aqueous solution of glucose is added per 60 ml of the autoclaved solution.
5. After spraying, the chromatograms are incubated in a moist atmosphere for 2-3 days at 25°C.
6. Inhibition zones indicate the presence of a fungitoxic product (or its conversion or decomposition products which are fungitoxic).

Alternative Method

It is easier and safer to prepare the agar plates containing bacteria or fungi and then placing the already developed chromatogram directly onto the agar for 10-15 min in order to let the test sample diffuse onto the agar. The main disadvantage in this procedure is, however, that some chemical compounds stick to the TLC plates tightly and may not get transferred to the agar.

1.3 ANTIVIRAL AND ANTICANCER ASSAYS

1.3.1 Anti-HIV Assay

AIDS (Acquired Immune Deficiency Syndrome) (**Caution!**)* is an immunosuppressive disease which has caused wide spread deaths through opportunistic infections and malignancies. A retrovirus, the human immunodeficiency virus (HIV), has been identified as the aetiologic agent causing this disease and approches to development of drugs against AIDS

***Caution** ! All manipulations of viral material should be performed in strictly contained environment (Laminar flow chamber) by using specially designed disposable gloves for virus handling. All contaminated materials and disposable items should be collected and burned in a controlled incinerator, away from urban areas.

are therefore based on finding substances which can inhibit HIV replication. There are several points at which the intervention of the replicative cycle can be carried out. The inhibition of a multifunctional enzyme, reverse transcriptase (which is a virus-specific RNA-dependent DNA polymerase) offers a possible mode of intervention of the virus life cycle since it transcribes the viral RNA genome to DNA which is ultimately incorporated as pro-viral DNA into the cellular genome. Recently a colorimetric test has been reported which is simple, sensitive and rapid involving the transformation of a tetrazolium salt to a coloured formazan derivative by living cells but not by dead cells or culture medium.

The procedure used at the National Cancer Institute, Bethesda, U.S.A. for testing activity against the Human Immunodeficiency Virus (HIV) is designed to detect agents acting at any stage of the virus reproductive cycle. The assay basically involves the killing of T4 lymphocytes by HIV (Schwarts, *et al.*, 1988). Small amounts of HIV are added to cells, and a complete cycle of virus, cells, or virus gene-products to interfere with viral activities is initiated to protect cells from cytolysis. The assay system is automated in order to accommodate a large number of test samples and is generally designed to detect anti-HIV activity. However, compounds that are decomposed or are rapidly metabolized in the culture conditions may not show activity in this bioassay screen. All tests are compared with atleast one positive (*e.g.* AZT-treated) control run simultaneously under identical conditions (Weislow *et al.*, 1989).

Materials

1. CO_2 incubator with temperature control
2. Tetrazolium salt, XTT
3. Dimethyl sulfoxide (DMSO)
4. T4 lymphocytes (CEM cell lines)
5. HIV-1 virus (extreme caution)[†]
6. Spectrophotometer
7. Compound microscope
8. 96-well plates
9. AZT
10. Multichannel pipettes
11. Test sample (crude extract, pure natural product or synthetic compound)[#]

[#] Only laboratories with trained technical staff, required facilities, incinerators and limited access areas should try to conduct this bioassay.

The anti-HIV assay consists of following steps:

1. The test sample is dissolved in dimethyl sulfoxide, then diluted 1:100 in cell culture medium before preparing serial half-$\log_{10}$ dilutions. T4 lymphocytes (CEM cell line) are added and after a brief interval HIV-1 is added resulting in a 1:200 final dilution of the compound. Uninfected cells with the compound (test sample) serve as a toxicity control, and infected and uninfected cells without the compound serve as basic controls.
2. Cultures are incubated at 37°C in a 5% carbon dioxide atmosphere for 6 days.
3. The tetrazolium salt, XTT, is added to all the wells, and cultures are incubated to allow formazan color development by viable cells.
4. Individual wells are analyzed spectrophotometrically for quantitative formazan production and, in addition, are viewed microscopically for detection of viable cells and confirmation of protective activity.
5. Drug treated virus-infected cells are compared with drug-treated non-infected cells and with other appropriate controls (untreated infected and untreated non-infected cells, drug-containing well without cells, etc.) on the same plate.
6. The data is reviewed in comparison with other tests done at the same time and the activity is determined.

1.3.2 Anti-Cancer Assays

A number of methods have been employed for the screening of anti-tumor compounds. The National Cancer Institute, Bethesda, Maryland, extensively used *in-vivo* testing of plant extracts against 3PS (P 388) (methylcholanthrene-induced) leukemia in mice and *in-vitro* screening for 9KB (human nasopharengeal carcinoma) cytotoxicity. However the *in-vitro* screening for 9KB cytotoxicity often leads to compounds which show no *in-vivo* activity, resulting in expensive dead-ends, since the *in-vitro* cytotoxicity does not correlate well with *in-vivo* activity. The *in vivo* 3PS test systems were therefore initially preferred, although they suffered from the disadvantages of being expensive (over US $ 200 per sample tested), time - consuming, complicated, sometimes not reproducible, and involved the use of laboratory animals. For these reasons the *in vivo* 3PS screen was eventually largely phased out and replaced by an *in vitro* screening system which comprised about 60 human cancer cell lines such as those of breast, colon, lung melanoma and other refractory solid tumors.

It is important to distinguish between the terms "cytotoxic", "antitumor" and "anticancer". "Cytotoxic" compounds are toxic to cells in culture but they may not show any selective toxicity to cancer cells as against normal cells. Cytotoxic compounds may be cytostatic (*i.e.* stop cell growth, reversibly or irreversibly) or cytocidal (kill cells).

"Antitumor" compounds are those which are active in an *in-vivo* tumor system. Such compounds would therefore show selectivity against tumor cells. "Anticancer" compounds are those which are effective in cancers in humans. Hence it requires human clinical trials to determine if any antitumor compound has anticancer activity.

1.3.2.1 In vitro anti-cancer screening

In the cancer field, *in vitro* assays are primarily of two types: molecular assays or cellular assays. Molecular assays are directed at a single subcellular target and they are therefore highly particular. They are of particular importance when a specific mechanism is of interest in a drug discovery program, and binding assays or inhibition assays can be used to discover new compounds having a specific type of activity. Because of their specificity, such assays result in a low hit rate from a large number of diverse samples screened. A battery of such screens is often used in conjunction in order to detect compounds working by more than one mechanism. A disadvantage is that interesting and important bioactive compounds not acting by the particular mechanism for which the screen is set up will be missed.

Cellular assays may be divided into two types: (a) cytotoxicity assays, and (b) other assays types (including morphological assays). A simple example of a cytotoxicity assay may be to measure the 50% growth inhibitory concentration against a single cell line, but this could lead to a large number of "active" materials, many of which could be uninteresting substances such as detergents, heavy metals, protein denaturants, non-selective DNA alkylating agents, mitochondrial poisons etc. Selecting out the really interesting compounds from the large number of "hits" can be difficult. The final choice of the type of assay to be employed must depend on the precise interest of the researcher. It is possible to use both types of assays in conjunction with one another, the initial cytotoxicity assays giving a large number of positive leads which are then further screened by biochemical assays to select compounds acting through mechanisms of interest. One must however take care of employing antagonistic screens *i.e.* the positive leads of one assay should be further tested in another assay which is working by an unrelated mechanism.*

Cell growth and cytotoxicity assays

There are four main types of non-radioactive cell growth and cytotoxicity assays:

i) cell or colony counts, and assays of
ii) macromolecular dye binding
iii) metabolic impairment and
iv) membrane integrity

No single method is universally appropriate for all situations. Each has limitations, and all are subjected to potentially serious artifacts under certain circumstances. Cell and colony counts are time consuming, tedious and sensitive to minor variations in methodology. Dye binding assays come closest to fulfilling the ideal requirements for growth and cytotoxicity assays. They are simple, rapid, reliable, sensitive and quantitative but do require access to ELISA reader. Metabolic impairment assays measure the decay of enzyme activity or metabolite concentration following toxic insult. They are generally more complex and artifact prone than dye binding assays. Membrane integrity assays measure the ability of cells to exclude impermeant extracellular molecules. They can either be colorimetric or fluorescent and require an ELISA reader and/or a fluorescent plate reader. They tend to be less artifact prone than metabolic impairment assays.

Experimental design

Seeding Density

Seeding density basically depends on cell size, growth rate and assay duration and must be determined individually for each cell type. However, in a 48-72 h. assay, seeding density is usually kept between 5 x 10^3 - 10^4 cells/well in 96 well microtiter plate.

Drug Solubilization

Stock solutions (1.0 mg/0.05 ml) of polar compounds are made in water while of non-polar compounds are made in 1:1 EtOH DMSO and the diluted with complete medium to the final test concentration. DMSO is

* For detailed protocols of anticancer screening see Daston, 1996; Fricker, P., *et al.*, 1996 and Basha, *et al.*, 1996.

toxic to most cells at concentrations above 0.5%. Preliminary experiments should be carried out to determine its toxicity threshold for each individual cell line.

Assay Duration

Assay duration depends on growth rate and seeding density and should be determined individually for each cell type. However for most transformed cell lines 48-72 h. period is usually adequate to detect the effect of drug.

Control Wells

Every test plate should include following five different types of samples:

i)	Medium blanks (MB)	growth medium with no cells or drugs
ii)	Drug blanks (DB)	growth medium with drug but no cells
iii)	-ve control	cells plus medium
iv)	+ve control	cells plus standard drug(s)
v)	Test	medium plus cells plus test compound

A. *Dye binding assay*

Sulforhodamine B (SRB) assay

Sulforhodamine B (SRB) is a bright pink aminoxanthene dye. Under mildly acidic conditions, SRB binds to basic amino acid residues of TCA fixed proteins. It provides a stable end-point that does not have to be measured within any fixed period of time. Once stained and air dried, plates can be kept for months before solubilization and reading. This assay has proven particularly useful in large scale drug screening.

Materials

1. Human tumor cell lines H157 and H1299 lung carcinoma, HT-144 malignant melanoma, Zr-75-1 and MCF7 breast carcinoma, SK-CO-1 and SW403 colon carcinoma, SK-OV-3 ovarian carcinoma and HT1376 bladder carcinoma
2. Tissue culture flasks 25 cm^2
3. 96 well microtiter plates
4. Medium RPMI-1640 buffered with 2.2 g l^{-1} $NaHCO_3$ supplemented with 10% heat inactivated foetal bovine serum (HIFBS); pH 7.4
5. DMSO-etOH 1:1
6. Double distd. deionized water

7. 10% Trichloroacetic acid (TCA)
8. 0.1% SRB in 1.0% glacial acetic acid
9. 10 mM unbuffered Tris base
10. CO_2 incubator
11. ELISA reader
12. Adriamycin, *cis*-Platin, 5-fluorouracil, mitomycin C and vinblastine
13. Test sample (plant extract, pure natural product or synthetic compounds)

The general startegic procedure for SRB assay is given below:

1. Cells are seeded onto 96-well microtiter plates at a concentration of 5×10^4 - 10^5 cells ml^{-1}, volume 200 µl/well.
2. Plates are incubated at 36.5°C in humidified CO_2 (10%) incubator for 24 h.
3. Old medium is removed and fresh medium is added.
4. 10 ul containing various concentrations of test compound is added whereas +ve and -ve control has standard drug(s) and no drug respectively.
5. Plates are incubated for next 48-72 h. at 36.5°C in humidified CO_2 (10%) incubator.
6. After incubation, medium is removed for the wells and 200 ul of 10% TCA is added. Plates are kept at 4°C for 30 minutes.
7. TCA is removed, plates are washed gently under tap water and air dried at room temperature.
8. 100 ul SRB reagent is added into each well and left for 15 minutes.
9. SRB is removed, wells are washed four times with 1.0% acetic acid and air dried.
10. Stain is solubilized with 0.2 ml 10mM unbuffered Tris base and absorbance is measured at 540 nm.
11. ED_{50} value of compounds possessed cytotoxic activity is calculated.

B. Cellular biomass assays

1. Propidium iodide (PI) assay

Thionin, Azure A, and Toluidine Blue O are biomass stains that approach the protein stains in sensitivity. Propidium iodide is a general biomass stain that binds to RNA, DNA, proteins and glycosaminoglycans.

Materials

1. Human tumor cell lines H157 and H1299 lung carcinoma, HT-144 malignant melanoma, Zr-75-1 and MCF7 breast carcinoma, SK-CO-1 and SW403 colon carcinoma, SK-OV-3 ovarian carcinoma and HT1376 bladder carcinoma
2. Tissue culture flasks 25 cm^2
3. 96 well microtiter plates
4. Medium RPMI-1640 buffered with 2.2 g l^{-1} $NaHCO_3$ supplemented with 10% heat inactivated foetal bovine serum (HIFBS); pH 7.4
5. DMSO-EtOH 1:1
6. Double distd. deionized water
7. 20% propidium iodide in distd. water (light sensitive)
8. CO_2 incubator
9. Fluorescent plate reader
10. Adriamycin, *cis*-Platin, 5-fluorouracil, mitomycin C and vinblastine
11. Test sample (plant extract, pure natural product or synthetic compounds)

The following procedure is used for the PI assay:

1. Cells are seeded onto 96-well microtiter plates at a concentration of 5×10^4 - 10^5 cells ml^{-1}, volume 200 μl/well.
2. Plates are incubated at 36.5 °C in humidified CO_2 (10%) incubator for 24 h.
3. Old medium is removed and fresh medium is added.
4. 10 ul containing various concentrations of test compound is added whereas +ve and -ve control has standard drug(s) and no drug respectively.
5. Plates are incubated for next 48-72 h. at 36.5°C in humidified CO_2 (10%) incubator.
6. Plates are kept at -30°C for 2-6 h. and thawed at 50°C for 15 minutes.
7. To each well, 50 ul of 20% PI stock solution is added so that final PI concentration is 400 μg/ml.
8. Plates are incubated in dark for 60 minutes at room temperature.
9. Fluorescence is read at 530/590-620 nm in a fluorescent plate reader and ED_{50} is calculated.

2. *Hoechst 33258 fluorescence assay*

Hoechst 33258 is a UV excited blue bisbenzimidazole dye which selectively intercalates into the A-T rich regions of DNA.

Materials

1. Human tumor cell lines H157 and H1299 lung carcinoma, HT-144 malignant melanoma, Zr-75-1 and MCF7 breast carcinoma, SK-CO-1 and SW403 colon carcinoma, SK-OV-3 ovarian carcinoma and HT1376 bladder carcinoma.
2. Tissue culture flask 25 cm^2
3. 96 well microtiter plates
4. Medium RPMI-1640 buffered with 2.2 g l^{-1} $NaHCO_3$ supplemented with 10% heat inactivated foetal bovine serum (HIFBS); pH 7.4
5. DMSO-EtOH 1:1
6. Double distd. deionized water
7. TNE Buffer (10 nM Tris, 1 mM EDTA, 2 M NaCl, pH 7.4)
8. 2% Hoechst 33258 in TNE buffer (light sensitive)
9. CO_2 incubator
10. Fluorescent plate reader
11. Adriamycin, *cis*-platin, 5-fluorouracil, mitomycin C and vinblastine
12. Test sample (plant extract, pure natural product or synthetic compounds)

The procedure used for the Hoechst 33258 fluorescence assay is given below:

1. Cells are seeded onto 96-well microtiter plates at a concentration of $5 \times 10^4 - 10^5$ ml^{-1}, volume 200 µl/well.
2. Plates are incubated at 36.5°C in humidified CO_2 (10%) incubator for 24 h.
3. Old medium is removed and fresh medium is added.
4. 10 ul containing various concentrations of test compound is added whereas +ve and -ve control has standfard drug(s) and no drug respectively.
5. Plates are incubated for next 48-72 h. at 36.5°C in humidified CO_2 (10%) incubator.
6. Medium is removed from the wells.
7. Plates are kept at -80°C for 1-2 h. and thawed at 50°C for 15 minutes.
8. To each well, 100 ul distd. water is added and plates are incubated at room temperature for 1 h.
9. Plates are refreezed at -80°C for 90 minutes and thawed at room temperature.
10. 0.1 ml of TNE containing 20 µl ml^{-1} of Hoechst 3325 dye is added and mixed well on a plate shaker.

11. Plates are incubated in dark for 90 minutes at room temperature.
12. Fluorescence is read at 350/460 nm in a fluorescent plate reader and ED_{50} is calculated.

C. *Metabolic impairment assays*

1. *MTT assay*

MTT assay measures the metabolic activity of living cells by their ability to reduce the tetrazolium salt, MTT, to form formazan using the dehydrogenase enzymes in the mitochondria.

Materials

1. Human tumor cell lines H157 and H1299 lung carcinoma, HT-144 malignant melanoma, Zr-75-1 and MCF7 breast carcinoma, SK-CO-1 and SW403 colon carcinoma, SK-Ov-3 ovarian carcinoma and HT1376 bladder carcinoma
2. Tissue culture flask, 25 cm^2
3. 96 well microtiter plates
4. Medium RPMI-1640 buffered with 2.2 g l^{-1} $NaHCO_3$ supplemented with 10% heat inactivated foetal bovine serum (HIFBS); pH 7.4
5. DMSO-EtOH 1:1
6. DMSO
7. Double distilled deionized water
8. 0.2% MTT in PBS (phosphate buffered saline)
9. CO^2 incubator
10. ELISA reader
11. Adriamycin, *cis*-platin, 5-fluorouracil, mitomycin C and vinblastine
12. Test sample (plant extract, pure natural product or synthetic compounds)

The systemic procedure for MTT assay is given below:

1. Cells are seeded onto 96-well microtiter plates at a concentration of 5×10^4 - 10^5 cells ml^{-1}, volume 200 µl/well.
2. Plates are incubated at 36.5°C in humidified CO_2 (10%) incubator for 24 h.
3. Old medium is removed and fresh medium is added.
4. 10 µl containing various concentrations of test compound is added whereas +ve and -ve control has standard drug(s) and no drug respectively.

5. Plates are incubated for next 48-72 h. at 36.5°C in humidified CO_2 (10%) incubator.
6. After incubation, medium is removed from the wells and 150 ul of fresh medium + 50 ul MTT is added. Plated are incubated for 4 h. at 36.5°C in humidified CO_2 (10%) incubator.
7. Medium/MTT is removed and insoluble formazan product is dissolved in 50 µl DMSO.
8. Absorbance is measured at 540 nm.
9. ED_{50} value of compounds possessed cytotoxic activity is calculated.

2. *Neutral Red assay*

Neutral Red is a vital dye that accumulates in the lysosomes of living cells. Dead and severely traumatized cells do not accumulate and retan Neutral Red.

Materials
1. Human tumor cell lines H157 and H1299 lung carcinoma, HT-144 malignant melanoma, Zr-75-1 and MCF7 breast carcinoma, SK-CO-1 and SW403 colon carcinoma, SK-OV-3 ovarian carcinoma and HT1376 bladder carcinoma.
2. Tissue culture flasks, 25 cm^2
3. 96 well microtiter plates
4. Medium RPMI-1640 buffered with 2.2 g 1-1 NaHCO3 supplemented with 10% heat inactivated foetal bovine serum (HIFBS); pH 7.4
5. DMSO-EtOH 1:1
6. Double distd. deionized water
7. 0.4% Neutral Red in distd. water (light sensitive)
8. 4.0% formaldehyde in 1.0% calcium chloride
9. Solubilization fluid; 1.0 ml glacial acetic acid in 100 ml of 50% EtOH
10. CO_2 incubator
11. ELISA reader
12. Adriamycin, *cis*-platin, 5-fluorouracil, mitomycin C and vinblastine
13. Test sample (plant extract, pure natural product or synthetic compounds)

The general procedure used for neutral red assay is given here:
1. Cells are seeded onto 96-well microtiter plates at a concentration of 5×10^4 - 10^5 cells ml^{-1}, volume 200 µl/well.

2. Plates are incubated at 36.5°C in humidified CO2 (10%) incubator for 24 h.
3. Old medium is removed and fresh medium is added.
4. 10 ul containing various concentrations of test compound is added whereas +ve and -ve control has standard drug(s) and no drug respectively.
5. Plates are incubated for next 48-72 h. at 36.5°C in humidified CO_2 (10%) incubator.
6. Within 24 h. of use, stock Neutral Red is diluted 1:80 with growth medium, pre warmed to 37°C and centrifuged at 700x g for 5 minutes to remove insoluble dye crystals.
7. Growth medium is removed from the plates and replaced with 0.2 ml of diluted neutral red solution per well.
8. Plates are incubated for 3-4 h. at 37°C.
9. Neutral Red is removed and wells are washed with 4% formaldehyde in 1% $CaCl_2$.
10. 0.2 ml of solubilization fluid is added and plates are solubilized for 15 minutes on a plate shaker.
11. Absorbance is recorded at 540 nm and ED_{50} is calculated.

D. Membrane integrity assay

Fluorescein diacetate assay

Fluorescein diacetate (FDA) is an electrically neutral non-fluorescent molecule. Viable cells accumulate FDA intracellularly and hydrolyze it to fluorescein. Fluorescein is retained intracellularly for a short period of time, causing them to become temporarily fluorescent.

Materials

1. Human tumor cell lines H157 and H1299 lung carcinoma, HT-144 malignant melanoma, Zr-75-1 and MCF7 breast carcinoma, SK-CO-1 and SW403 colon cercinoma, SK-OV-3 ovarian carcinoma and HT1376 bladder carcinoma
2. Tissue culture flasks, 25 cm^2
3. 96 well microtiter plates
4. Medium RPMI-1640 buffered with 2.2 g l^{-1} $NaHCO_3$ supplemented with 10% heat inactivated foetal bovine serum (HIFBS); pH 7.4
5. DMSO-EtOH 1:1
6. Double distilled deionized water

7. 1.0 mg/100 ml FDA in DMSO (distributed in aliquots and stored at -20⁴C)
8. Adriamycin, *cis*-platin, 5-fluorouracil, mitomycin C and vinblastine
9. Test sample (plant extract, pure natural product or synthetic compounds)
10. Fluorescent plate reader

The fluorescein diacetate assay procedure is given below:

1. Cells are seeded onto 96 well microtiter plates at a concentration of $5 \times 10^4 - 10^5$ cells ml^{-1}, volume 200 µl/well.
2. Plates are incubated at 36.5ºC in humidified CO2 (10%) incubator for 24 h.
3. Old medium is removed and fresh medium is added.
4. 10 ul containing various concentrations of test compound is added whereas +ve and -ve control has standard drug(s) and no drug respectively.
5. Plates are incubated for next 48-72 h. at 36.5ºC in humidified CO_2 (10%) incubator.
6. Medium is removed by inverting plates and flicking gently and washed once with PBS.
7. 200 µl of prewarmed FDA is added into each well and plates are incubated at 37ºC for 60 minutes.
8. Plates are centrifuged at 200x g for 5 minutes and solution is removed.
9. 200 µl of prewarmed PBS is added to each well and fluorescence is read immediately at 485/538 nm.
10. ED_{50} is calculated.

1.3.3 HCT Cytotoxicity Assay

The HCT116 cytotoxicity assay serves as the *in vitro*, primary biological screen that could be used for rapid cytotoxicity testing during the natural product isolation process. The cytotoxic activity is determined by incubating the cells with test sample for 48 hours in 96-well microtiter plates, and measuring viable cell number using the crystal violet staining technique. The cell line HCT116 (human Colon Tumor Cell) has been routinely used in the screening of natural products, including those derived from marine organisms. It has previously been reported that this *in vitro* assay is an excellent representative for *in vivo* activity and it is therefore the assay of choice during bioassay-guided fractionation.

Materials

1. Media
   - 1000 ml McCoys 5A Medium
   - 114 ml Fetal Bovine Serum
   - 25 ml Hepes buffer solution (1M)
   - 12.5 ml Penicillin-Streptomycin solution
2. Cell line: HCT116, human colon cancer cell line†
3. Fixing solution
   - 10% of 37% Formaldehyde
   - 10% Phosphate Buffered Saline 10 X
   - 80% distilled water
4. Crystal violet-0.0075% working solution
   - 62.5 ml 0.02% crystal violet
   - 104.17 ml distilled water
5. EDTA-Trypsin
6. Standard VP-16
   - Stock is 1 mg/ml in DMSO
   - IC_{50} is 0.7 μg/ml
7. Test sample: starting concentration is 10 mg/ml
8. 96-well flat-bottomed microtiter plate
9. Hemacytometer
10. Light microscope

The general procedure for the HCT cytotoxicity assay is given below:

1. Remove medium from the cells and add 2 ml of EDTA/Trypsin (enough to cover cells) and swirl. Trypsin digests proteins and detaches cells from the flask.
2. Remove trypsin, add 2-3 ml more trypsin swirl and remove.
3. Allow to stand in incubator for maximum five minutes. Allow cells to become completely detached from the flask.
4. "Slam" flask on bench top two to three times and hold vertical. Cell should flow down in the flask.
5. Add 10 ml of media. Should turn cloudy.
6. Mix 5 to 10 times with a pipet to break up clumps.
7. Look at under microscope to make sure you have single cells.
8. Use hemacytometer to count and find concentration of cells. Dilute cells if necessary.

---

†Available from ATCC

Use of Hemacytometer

- Count number of cells in each horizontal line of larger squares (five to a side of a grid) from left to right.
- If cell is on line only count those on right and bottom side.
- Add up total for the five rows.
- Grid is 1 mm squared. Depth is 0.1 mm. Therefore volume is 0.1 mm cubed or 0.1 μl.
- Concentration per ml is total counted times 10^4. Example: if count 85 cells have 0.85×10^6 cells/ml.

9. Determine number of plates needed. Multiply by 15 ml for total volume of cell suspension needed.
10. Dilute cells to final concentration of 25×10^3 cells/ml and final volume determined in step 9 in sterile flask.
11. Pour cells suspension into sterile trough.
12. Add 150 μl of cell suspension to each well.
13. Put plates in incubator overnight.
14. Add 50 μl of test compound at 2 mg/ml to first well. Add standard at stock solution.
15. Set 12 chamber pipet at 50 μl.
16. Remove 50 μlfrom first well and add to next well and mix. (makes a 1:4).
17. Repeat all the way down row. Discard the 50 μl from the last well.
18. After incubation for 3 days, remove lid, turn upside down and shake in order to remove media.
19. Place on paper towel upside down.
20. Prepare 500 ml of fix solution.
21. Add 10 μl of fix solution to each well.
22. Fix for 15 minutes.
23. Discard fix solution.
24. Invert to air dry.
25. When dry stain plates with 55 μl of 0.0075% crystal violet for 15 minutes.
26. Rinse plates twice with distilled water, invert and allow to air dry.
27. IC_{50} is point where cells begin to die. Color changes from solid blue to clear.

1.4 ANTIMITOTIC ASSAY

1.4.1 Antimitotic Assay Using Sea Urchin Eggs

Inhibition of cell division is a measure of the antimitotic activity of chemical compounds. Antimitotic chemical compounds such as vinblastine and podophyllotoxin have been shown to inhibit cell division of fertilized sea urchin eggs and starfish oocytes. The following bioassay provides an easy method of detecting the antimitotic activity of chemical compounds (Jacobs *et al.*, 1980, White *et al.*, 1981).

Materials

1. Male and female sea urchins (*Strongylocentrotus purpuratus*)*
2. KCl, 0.5-0.6 M
3. Light microscope
4. Sea water (made by adding 3.8 g of sea salt per liter of dist. water)+
5. Small vials
6. Incubator
7. Test tubes
8. Syringe
9. Test sample (crude extract, pure natural product or synthetic compounds)

The sequential steps involved in antimitotic assay on sea urchin eggs are as follows:

1. Sexually mature male and female sea urchins are induced to spawn by the injection of a small amount of KCl solution.
2. White sperms are collected from the male and kept in a small test tube at ice temperature.
3. The eggs collected from the female urchins are washed with cold sea water, and resuspened in sea water (400 ml) to produce a slurry.
4. Sperm (1-2 drops) is added to 50 ml sea water and 1 ml of this solution is added to the slurry of eggs for fertilization to occur.
5. Aliquots of the mixture are treated with different concentrations (16-50 mg/ml) of the test sample within five minutes after fertilization. If the test sample is not soluble in water, some organic solvent such as propylene glycol in microliter quantities can be used. An equal quantity of solvent should be added in the control vial.
6. The embryos are allowed to proceed to the first cleavage by placing them on ice after 2-3 hr. incubation.

---

* Information on the care and handling of sea urchin eggs, embryos and adults are available in the book "*The Sea Urchin Embryo*" [G. Czihak, ed.) New York, Springer-Verlag, p. 10-25 (1975)].

+ Available from Instant Ocean, Inc., 8141 Tyler Boulevard, Mentor, OH 44060, U.S.A.

7. The incubations are carried out at 14° or 15 °C with frequent agitation of the cells to minimize settling, promote contact inhibition and to ensure efficient sample distribution.
8. Inhibition of cleavage can be observed from random populations totaling 500-600 eggs under a light microscope after an incubation time of 2-3 hr. If 80 to 100 % inhibition of cleavage occurred at ~ 16 mg/ml, the compound is considered to be active.
9. The results are expressed as a percentage (the number of cells cleaved divided by the number of cells not cleaved) relative to a solvent treated control.

1.5 GENOTOXICITY ASSAYS

1.5.1 SOS Chromtest for Genotoxicity

The biochemical induction assay (BIA) involves the monitoring of the lysogenic induction of a lambda-*lac z* fusion phage by using *Escherichia coli* (strain BR 513). The *lac z* gene of this organism, which is responsible for the production of β-glycosidase, is fused to the promoter of phage lambda, and the expression of this gene causes what is known as an "SOS" response and results in the production of β-glycosidase which is measured colorimetrically.

In a modification of this assay, known as the "SOS chromtest", the expression of an SOS regulated gene (*sfiA* gene), which has been fused to the *lac z* gene, is monitored. A different strain of *Escherichia coli* (strain PQ 37) is employed. SOS chromtest using *Escherichia coli* K-12 is therefore also a colorimetric assay for detecting genotoxicity of chemicals. This is a modification of the original liquid suspension assay and involves a simple, convenient agar spot test. Since genotoxic agents also induce the production of β-galactosidase in strain PQ37, it can be detected colorimetrically. Hence, it provides a simple means of screening for potential chemical carcinogens or genotoxic agents that induce cell filamentation (Mamber *et al.*, 1986).

Materials

1. Tester strain: *E. coli* PQ37$^+$ (**Caution !**)
2. LB Broth

Difco-Bacto tryptone	10 g/l
Yeast extract	5 g/l
NaCl	10 g/l

	Trizma base (Sigma)	1 M
	(121.1 g/l)	6.5 ml
3.	Autoclave	
4.	Low temperature freezer (-80 °C)	
5.	Difco-Bacto agar	
	LB Agar	15 g/l
	Soft agar`	10 g/l
6.	3 N NaOH solution	
7.	Deionized water	
8.	Laminar flow chamber	
9.	Micropipettes with disposable tips	
10.	Colorimetric indicator-Fast Blue RR salt (Sigma)	65 mg
11.	6-Bromo-2-naphthyl-β-D-galactopyranoside (BNG,Sigma)	20 mg
12.	Dimethyl sulfoxide (DMSO)	1.5 ml
13.	S9 mix (9.1 ml), sodium phosphate buffer (0.2 M, pH 7.5, 7.5 ml), glucose 6-phosphate (0.1 M, 1.0 ml), NADP (0.1 M, 0.3 ml), $MgCl_2$ (0.4 M, 0.3 ml), filter sterilize, store at 4 °C	
14.	Bio-assay dishes (Nunc)	
15.	Vials	
16.	Colorimeter	
17.	S9 mix	9.1 ml
18.	Test sample (crude extract, pure natural product or synthetic compounds)	

The genotoxicity assay involves following steps:

Method of Preparing Microbiological Growth Medium

LB Broth with 15 g of Difco-Bacto agar are added per liter of deionized water and pH adjusted to 7.8 with 3*N* NaOH. The ingredients for LB agar are boiled and autoclaved at 121° C for 20 min, cooled and stored at 27-29 °C.

Method of Maintenance and Preservation of Strain

The *E. coli* PQ37 culture is inoculated into 100 ml LB broth and incubated with aeration at 37 °C for 18 hr. The strain should be stored as a frozen culture by mixing 8 ml of a broth culture with 0.7 ml sterile DMSO and dispensing into a sterile vial, which is then stored at -20 °C or -80 °C.

Chromtest Method

1. Strain PQ37 is inoculated from a frozen vial into LB broth (15 ml in a 50 ml conical flask) and incubated on a shaker at 37 °C for 18 h.
2. 2 ml of the overnight culture is added to a flask containing 30 ml of molten soft agar (40-45 °C), mixed and 30 ml are poured onto the surface of a pre-warmed bio-assay dish, 243 mm x 243 mm, containing 50 ml of LB agar.
3. The overlay is spread evenly across the LB agar base and allowed to harden at room temperature. Dishes can be refrigerated for use for upto 6 hr. after preparation.
4. When metabolic activation is required, 2 ml of the overnight culture is added to 19 ml of the molten soft agar (40-45 °C) followed by further addition of 9 ml of S9 mix. This overlay is immediately poured onto a bio-assay dish, spread and allowed to harden at room temperature. Dishes containing the S9 mix are refrigerated and used within 1 hr. of pouring.
5. Using micropipettes with disposable tips, 20 ml of the test sample is directly spotted on the agar surface.*
6. Immediately after dispensing the sample, dishes are incubated at 37 °C for a minimum of 16 h (16-24 h).
7. A colorimetric indicator overlay is prepared according to the following formula:

Fast Blue RR salt (Sigma)	65 mg
6-Bromo-2-naphthyl-β-D-galactopyranoside (BNG, Sigma)	20 mg
DMSO	1.5 ml
Soft agar	25 ml

8. 29.5 ml of the colorimetric indicator is poured over each bio-assay dish and spread evenly.
9. Dishes are left at room temperature for 10-20 min to allow for agar hardening. The development of a dark red color reaction surrounding the site of sample application is an indication of β-galactosidase induction. Scoring of the β-galactosidase activity is based upon the relative intensity of the color reaction (0 =

---

* In qualitative tests, a single relatively large (1-10 mg/ ml) concentration of the test sample is used, while in semi-quantitative tests multiple concentrations (100, 30, 10, 3, 1 mg/ml) are employed.

negative, +1= equivocal, +2 = weak positive, +3 = moderate positive, +4 = strong positive).

10. 4-Nitroquinoline-*N*-oxide (10 mg/ml), streptonigrin (10 mg/ml) or mitomycin C can be used as positive controls.

1.5.2 Ames Test for Detecting Carcinogens and Mutagens

Since 1975, many rapid bacterial assays for detecting carcinogens and mutagens have been proposed. As a result it is now clear that there is a strong correlation between the ability of chemicals to be genotoxic to bacteria and their mutagenic properties in mammals. Since bacteria grow rapidly in simple defined media, bacterial tests are among the simplest, quickest and least expensive to conduct.

The *Ames test,* also called the *Salmonella*/Microsome assay, is among the best short term bacterial tests. This assay measures reverse mutations from histidine auxotrophy to prototrophy in several specially constructed mutants of *Salmonella typhimurium* by a wide variety of mutagens. Therefore the test has been used to detect chemicals which are potentially carcinogenic or mutagenic to exposed humans.*

Materials

1. Bacterial tester strains:
   A set of histidine requiring bacterial strains are used, originally derived from *S. typhimurium* LT2 e.g strain TA 1535, TA 100, TA 1537 and TA 98 (**Caution !**)#
2. Microscope
3. Automatic colony counter
4. Gyrorotory incubator
5. Stationary incubator
6. Laminar flow hood
7. Freezer (-80 °C)
8. Adjustable micropipettes with disposable tips
9. 1 dram c/ml glass vials
10. Disposable glass culture tubes (13 x 100 mm)
11. Latex surgical gloves
12. Sterile 1/4 inch filter paper discs
13. Plastic cryotubes

---

* For further details about Ames test for detecting carcinogens and mutagens, see Ames *et al.*, 1973; Maron *et al.*, 1973 and Belser *et al.*, 1981.

#Caution!: pathogenic

14. Sealed filteration units (0.2 μ pore size)
15. Vortex
16. *Minimal agar*: (Vogel-Bonner medium E)

$MgSO_4 . 7H_2O$	10 gm
Citric acid monohydrate	100 gm
K_2HPO_4	500 gm
$NaHNH_4 PO_4.4H_2O$	175 gm
Warmed distilled water (45 °C)	670 ml

(Adjust the volume to 1 l. by adding distd. H_2O)

17. *0.5 mM Histidine/Biotin Solution*

D-Biotin	30.9 gm
L-Histidine. HCl	24 gm
Distd. H_2O	250 ml

18. *Top agar*

Agar	6 gm
NaCl	5 gm
Distd. H_2O	1000 ml

19. *S9 Mix*

Rat liver (Aroclor induced)	2.4 ml (4%)
$MgCl_2$-KCl salts	1.0 ml
1M Glucose-6-phosphate	0.25 ml
0.1 M NADP	2.0 ml
0.2 M Phosphate buffer (pH 7.4)	25.0 ml
Sterile distilled H_2O	19.75 ml

20. *Minimal glucose agar media*

Agar	15 gm
50 x VB salt	20 ml
40% Glucose	50 ml
Distd. H_2O	340 ml

The Ames test can be employed by using the following method:

A. Method of Preparing Microbiological Growth Media Used

Top agar :

1. Top agar, containing 0.6% Difco agar and 0.5% NaCl, is autoclaved and stored at room temperature in a volume of 100 ml in screw-capped tubes.
2. Before use the top agar is melted and 10 ml of sterile solution of 0.5 mM L-histidine HCl/ and 0.5 mM biotin are added.
3. A trace amount of histidine in the top agar allows all the bacteria on the plate to undergo several multiplications and produces a

faint background bacterial lawn which is visible to the naked eye and can be examined under a dissecting microscope.

Agar plates:

1. Plates for the mutagenicity assay contain 30 ml of minimal glucose agar medium. Sterile, disposable plastic Petri plates (100 mm x 15 mm) are used for this purpose.

B. Method of Maintenance and Preservation of Strain

A culture of *Salmonella typhimurium* strain is inoculated into Oxoid nutrient broth and incubated at 37 °C for 24 hr. For each 1.0 ml of culture, 0.09 ml of sterile DMSO is added and the culture aseptically distributed into sterile 1.2 ml cryotubes, which are then stored at -80 °C.

C. Plate Incorporation Method

1. Tester strains are reisolated from the frozen master copies by streaking the bacteria on minimal glucose agar plates enriched with histidine and biotin. Plates are incubated for 24 hrs at 37 °C with a well isolated colony picked by a sterile wire loop for overnight growth in Oxoid nutrient broth.
2. 0.1 ml of a fresh overnight culture of the tester strain is added to a 13 x 10 mm capped culture tube containing 2 ml of top agar, and kept at 45 °C in a heating block.
3. 0.1 ml or less of test sample is added along with 0.5 ml S9 Mix.
4. The test sample is tested with and without the S9 Mix and both positive and negative control plates are included in the assay. Negative controls contain the bacteria, S9 Mix, and solvent (but not the test sample) while the positive control contains standard diagnostic mutagens e.g. streptonigrin, mitomycin, etc.
5. The test sample is mixed by vortexing the soft agar for 3 sec. at low speed and then poured onto glucose agar plates, tilted and the plates quickly rotated and then allowed to harden at room temperature (the mixing, pouring and distribution should not take more than 20 sec).
6. The plates are covered with brown paper to avoid the effect of light on photosensitive chemicals.
7. Within an hour the plates should be inverted and placed in a dark and vented incubator at 37 °C.

8. After 48 hr. the *revertant* colonies on test plates and on controls are counted, and the presence of the background lawn on all plates is confirmed.
9. A lawn that is thin compared to the lawn on the negative control plates is evidence of bacterial toxicity. The number of revertants per plates depends on the number of histidine auxotrophs in the lawn after 48 hrs. incubation (- =<20; + => 20; ++ => 100; +++ => 500 revertants/plate).

Spot test

This is a variation of the plate incorporation test in which the mutagen is directly applied to the surface of the minimal agar plates after it has been seeded with the bacterial tester strain and S9 Mix. The spot test is simple and eliminates the time-consuming preparation of solutions of chemicals to be tested, but it has some limitations, such as:
(a) it can be used only for testing chemicals that are diffusable in the agar,
(b) it is much less sensitive than the standard plate incorporation test.

1.5.3 Antiphage Activity Assay

Bacterial viruses can be used as a convenient model for the pre-screening of antiviral and antineoplastic activity. Antiphage screening is an inexpensive, rapid and sensitive bioassay procedure (Delitheos, 1992).

Materials
1. Host strain *E. coli* B (ATCC 11303) for bacteriophages T2 (ATCC 11303-B2), T3 (ATCC 11303-B3), T4 (ATCC 11303-B4), T7 (ATCC 11303-B7), (**Caution !**)*
2. *E. coli* C (ATCC 13706) for phage øX174 (ATCC 13706-B1) (**Caution !**)*
3. *Pseudomonas aeruginosa* 7 for ø Ps 7 (**Caution !**)*
4. Tryptone soya agar (TSA - Oxoid)
5. Tryptone soya broth (TSB-Oxoid)
6. Petri dishes
7. Pasteur pipette
8. Incubator

*Caution! Pathogenic

9. Test sample (plant extract, pure natural product or synthetic compound)

The systemic procedure for antiphage activity assay is as follows:

1. The bacterial culture of each host strain is prepared by inoculation of 5 ml TSB with 1-2 bacterial colonies.
2. After 1 hr incubation at 37 °C, TSA petri dishes are flooded with the appropriate amount of this culture.
3. The excess broth is carefully removed with a Pasteur pipette and the plates are allowed to dry in an incubator for 1 hr.
4. Aliquots (0.5 ml) of each phage tested, diluted at the corresponding RTD (Routine Test Dilution), are transferred into test tubes.
5. 0.5 ml of the test sample is added into the test tube and the mixture is allowed to stand at room temperature for 30 min.
6. One drop of each mixture is placed onto the surface of the agar seeded with the host strain. After a few minutes, the drop is absorbed by the agar on the petri plate.
7. These plates are incubated overnight at 37 °C.
8. Antiphage activity of the extract is noted by the absence of plaque formation in comparison with controls where typical bacterial lysis occurs.

1.6 ASSAYS FOR CONTROL OF TROPICAL DISEASES

1.6.1 Antimalarial Assays

There are four species of the parasite *Plasmodium* which are responsible for causing malaria in humans. They are *P. falciparum, P. vivax, P. malariae* and *P. ovale*. Female anopheline mosquitoes infected with *Plasmodium* sporozoites transfer the *Plasmodium* sporozoites to the blood of humans by bites. In the initial stage (the pre-erythrocytic phase), the sporozoites disappear from the blood and invade the liver. The malaria symptoms are not apparent at this stage. Merozoites released from the liver then invade the erythrocytes, thereby starting the blood cycle. The merozoites in the erythrocytes are transformed into trophozoites. On asexual reproduction of trophozoites, schizoites are formed which burst, thereby affording many more merozoites which again infect more erythrocytes, thereby multiplying the number of infected erythrocytes. When the erythrocytes burst, an intense feeling of cold is produced in the patient, causing shivering. Some of the merozoites then become differentiated into male and female gametocytes which cannot develop further in the human being and must be transferred to the female *Anopheles* mosquitoes gut where the sexual phase of the life cycle of the malarial parasites takes its course [Gupta, 1995].

1.6.1.1 96 Well plate in vitro schizonticidal assay

Materials

1. Malarial parasite (*P. falciparum*) culture
2. 96 well microtiter plates
3. Medium RPMI-1640 buffered with 40 mM TES and supplemented with 2.0 g l^{-1} glucose and 10% human O+ serum; pH 7.4
4. Fresh human RBCs
5. Double distilled deionized water
6. Dimethyl sulfoxide (DMSO)
7. Absolute methanol
8. Giemsa stain
9. CO_2 incubator
10. Light microscope
11. Chloroquine
12. Test sample (plant extract, pure natural product or synthetic compounds)

The procedure adopted for antimalarial assay is as follows:

1. A drop of the malarial parasite culture is taken and at least 500 erythrocytes are counted, making a note of the number that contain parasites (excluding gametocytes). To estimate the percentage parasitemia, the number of infected erythrocytes is divided by 5.
2. The culture is diluted so that final parasitemia is 2.0%.
3. The culture is grown at 37°C in the presence of 5% CO_2 for 32 h. The erythrocytes are centrifuged and resuspended in fresh culture medium RPMI-1640, supplemented with 10% human O+ serum.
4. 100 μl aliquots are distributed into sterile 96 well microtiter plate and 10 μl containing various concentrations of the test compound (solubilized in 0.5% DMSO/MeOH) is added and culture is placed in humidified CO_2 (5%) incubator at 37°C.
5. -ve control is administered 10 μl PBS in place of the drug while +ve control contains standard drug(s).

6. Thin blood films of the culture are prepared after 24, 36 and 72 h. The slides are stained with Giemsa stain and the number of parasites per 1000 erythrocytes are determined by microscopic examination.
7. Total parasitemia is calculated and plotted as percentage control against concentration and IC_{50} is determined.

1.6.1.2 Suppressive Test of Blood Schizonticidal Action (Peter's test)

Materials

1. Male mice (Swiss Albino)
2. Tissue culture medium (TIC 199)
3. Syringe
4. Sonicator
5. Tween 80
6. Carbomethoxycellulose (CMC)
7. Dimethyl sulfoxide (DMSO)
8. Staining agent (Gremisa)
9. Microscope
10. Test sample (crude extract, pure natural product or synthetic compound)

The following procedure is employed for Peter's test:

1. This test requires male mice kept at 22 ^{o}C (± 2 ^{o}C) in batches of five.
2. Blood is withdrawn from the donor infected mouse (with approx. 20% infected erythrocytes) which is diluted in tissue culture medium so that each 0.2 ml contains about 10^7 infected cells.
3. Each mouse is infected with 0.2 ml of this infected blood on the first day by intravenous injection from the tail.
4. The test sample (suspended or dissolved by trituration/sonication in 2% aqueous solution of Tween 80 or 0.5% carbomethoxycellulose or dimethyl sulphoxide) is administrated either orally or subcutaneously (dose 1-100 mg per kg on successive days) on the 5th day.
5. Samples taken from tail blood are stained (e.g. with Gremisa) and the extent of infection is recorded (parasitized cells as a percentage of total cells).
6. ED_{50} values (50% suppression of parasites as compared to controls) are calculated.

7. For preliminary large scale screening a single dose of test sample (e.g. 100 mg/kg) instead of daily doses may be given.

1.6.1.3 Rane Test of Blood Schizonticidal Activity

Materials

1. Male Mice (Swiss Albino)
2. Malaria Infected mouse
3. Syringe
4. Test sample (crude extract, pure natural products or synthetic compounds).

The Rane test of blood schizonticidal activity follows this procedure: This test involves giving a total dose of the standard inoculum of *Plasmodium berghei* which kills mice within 6 days in the presence or absence of a single dose of the test sample, and comparing survival times. An extension of survival times 12 days or longer is considered significant.

1. The mice are infected with a dose of 10^6 infected donor cells on the first day and the test sample is given subcutaneously at an initial dose range of 600, 300, 150 and 75 mg per kg.
2. Survival times for test and control mice is noted.

1.6.2 Larvicidal Assay

Malaria, yellow fever and dengue fever are some of the most widespread parasitic diseases prevalent in the tropics and sub-tropics. Mosquitoes of the genera *Anopheles* and *Aedes* are vectors of these diseases. One way of controlling tropical diseases is by eliminating the vector of the parasite. Plant-derived and other natural larvicides can be used for the elimination of these vectors (Zarroug *et al.*, 1988).

Materials

1. Common mosquitoes such *Anopheles arabiensis* (vector of malaria), *Aedes aegypti* (vector of yellow fever) and *Culex quinquefasciatus* (urban nuisance mosquito).
2. Temperature controlled insectary
3. Turf (a piece of soil with grass and plant roots)
4. Baby food such as Farex
5. Rabbits

6. Small quantities of fresh human blood (**Caution**!)†
8. Glucose
9. Petri dishes
10. Filter paper
11. Distd. water
12. Small bowls

The larvicidal assay involves following steps:

1. Mosquitoes are maintained in the insectary with a temperature range of 28-30 °C and 70-80% relative humidity. A piece of turf is added as a source of food for young larvae.
2. After molting to the second instar stage, the larvae are fed with a small amount of baby food. Adults are fed on rabbits. Female *An. arabiensis* are occasionally fed with human blood. Male *An. arabiensis* are fed on a 10% glucose solution.
3. Mosquito eggs are collected on moist filter paper placed on a petri dish. These eggs can be kept in a regulated atmosphere (26-28 °C, 70-80% rel. humidity) for upto six months.
4. The larvae hatch readily when the eggs are placed in a bowl of tap water.
5. Different concentrations of test sample (1000, 500, 250, 100, 50, 10 ppm) are dissolved in distd. water (small bowls) for the test. A minimum of eight replicates are required for each concentration.
6. One-day old larvae* (25 per bowl) are then transferred into bowls containing the test sample. The exposure period is 24 h during which no food is offered to the larvae.
7. The killing effect of the test sample is assessed after 30 min. and 24 hr of exposure.
8. Percent mortality is calculated to represent the larvicidal activity of each sample.

1.6.3 Molluscicidal Activity

The tropical endemic disease, schistosomiasis, is spreading rapidly and over 200 million people are infected with this disease. The snail *Biomphalaria glabrata* acts as the vector for the disease. Vector control is always considered to be an important step in the irradiation of tropical diseases. The molluscicidal assay provides a simple tool to screen

---

†**Caution !** All safety regulations and precautions should be followed.

* Second and fourth instar larvae

different natural products for the control of vector snails (Hostettmann *et al.*, 1982).

Materials

1. Snails of the species *Biomphalaria glabrata* kept in acquaria with a continuous circulation of water through a filtering system (water temperature at 24 °C)
2. Distd. water
3. Petri dishes
4. Microscope
5. Ultrasonic bath
6. Test sample (crude extract, pure natural product or synthetic compound)

The foollowing steps are involved in the molluscicidal activity assay:

1. Solutions of the test sample in distilled water (32, 16 , 8, 4, 2 ppm) are prepared with two replicates for each treatment. If the compounds or extracts are weakly soluble in water, solutions should be placed in an ultrasonic bath for one hour prior to the bioassay.
2. Two snails of uniform size (average diameter of the shell, 9 mm) are placed in each concentration of the test samples.
3. After 24 hours these snails are transferred to Petri dishes, light is shone from the bottom of the Petri dish and the heart-beat is checked by a microscope.
4. Mortality is counted at each concentration and the result of the bioassay is recorded as + ve (active) or -ve (inactive).

1.6.4 Piscicidal Assay

There is a close relationship between molluscicidal and piscicidal activities, since the molluscicidal compounds are also usually very toxic to fish. Piscicidal compounds often possess other bioactivities such as insecticidal, plant growth inhibitory, insect antifeedant, antitumor , co-carcinogenic and irritant activities (Kawazu, 1981).

Materials

1. Killie-fish (*Oryzias latipes*)[+]
2. Organic solvents (methanol, acetone, DMSO, etc.)
3. Air pump
4. Pipettes

[+] Commercially available as an ornamental fish.

5. Beakers (200 ml)
6. Incubator
7. Water tank
8. Fish food
9. Test sample (crude extract, pure natural product or synthetic compound)

The piscicidal activity assay involves the following steps:
1. Killie-fish averaging 350 mg in weight and 3-3.5 cm in length are reared in a water tank and not fed for two days, prior to the test.
2. Various concentrations of test sample in organic solvents such as methanol (5000, 2500, 1000, 500 ppm) are prepared.
3. These solutions are then added to beakers containing 150 ml water and aerated for two days with an air pump before the test.
4. Organic solvent alone serves as a control.
5. Five killie-fish are introduced into beakers containing various concentrations of test sample. These beakers are then kept in an incubator to maintain the temperature between 18-19 ^{o}C.
6. Fish that die during the test are immediately removed from the solutions to avoid toxic effects.
7. The number of surviving fish in each beaker is recorded after 24 h after the initial introduction.
8. The results are represented as minimum lethal concentrations (MLC).
9. If more than one fish dies in the control beaker, the test should be repeated.
10. If test sample is poorly soluble in methanol, the minimum possible quantity of acetone or DMSO can be used. The same quantity of this solvent should be mixed into the control.

1.6.5 Amoebicidal Assay

The pathogenic protozoa, *Entamoeba histolytica*, causes amoebiasis in humans. Health disorders such as amoebic dysentery and local necroses (abscesses) in organs including liver, brain and lung are caused by this condition. Amoebicidal compounds from natural sources can be studied for their potential to control the disease (Keene, 1986).

Materials
1. *Entamoeba histolytica* (NIH 200 strain) (**Caution!**)*
2. DMSO

***Caution** ! Pathogenic.

3. 0.22 mm filters
4. Autoclave
5. Petri dishes
6. Inactivated horse serum
7. NaOH solution (1 *M*)
8. Falcon flasks (Nunc, 75 and 25 cm^2)
9. Haemocytometer
10. Inverted microscope
11. Basic TPS 1 medium (Diamond's medium) containing the following chemicals in 900 ml distd. water:

Trypticase	10 g
Panmede (Ox liver digest)	20 g
Glucose	5 g
L-Cysteine hydrochloride	1 g
Ascorbic acid	0.2 g
NaCl	5 g
Potassium dihydrogen orthophosphate	0.6 g
Dipotassium hydrogen orthophosphate	1

12. Vitamin mixture:
The mixture is prepared by adding water soluble vitamin B 500 ml (I), biotin solution 250 ml (II), folic acid solution 250 ml (III), lipid soluble vitamin 2,500 ml (IV), and vitamin E solution 250 ml (V). The solution is sterilized by filtration through 0.22 mm disposable filter and stored at 4 °C.

(I) ***Vitamin B solution***:

Solution *a*	
Niacin	62.5 mg
p-Aminobenzoic acid	125 mg
Distd. water	150 ml
The solution is cooled and made upto volume with distd. water	150 ml
Solution *b*	
Niacinamide	62.5 mg
Pyridoxine hydrochloride	25 mg
Pyridoxal hydrochloride	62.5 mg
Thiamine hydrochloride	62.5 mg
Calcium panthothenate	25 mg
i-Inositol	125 mg
Choline chloride	1,250 mg
Distd. water	upto 150 ml

Solution *c*

Riboflavin	25 mg
Distd. water	75 ml
0.1 *N* NaOH	drop-wise to dissolve

Volume is made upto 100 ml with distd. water.

(II) ***Biotin solution***:

D-Biotin	200 mg
Distd. water	75 ml
0.1 *N* NaOH	dropwise to dissolve

Volume made upto 300 ml. with distd. water

(III) ***Folic acid solution***:

Folic acid	30 mg
Distd. water	upto 200 ml
0.1 *N* NaOH	dropwise to dissolve

Volume is made up to 300 ml. with distd. water

(IV). ***Lipid soluble vitamins solution***: The following two solutions *a* and *b* are combined together and made up to 3000 ml with dist. water.

Solution *a*

Vitamin D_2 (calciferol)	300 mg
Crystalline alcohol of vitamin A	300 mg
95% Ethanol	63 ml

Solution *b*

Vitamin K (menadione sodium disulfite)	60 mg
5% v/v Tween 80	upto 100 ml

(V). ***Vitamin E solution***:

Vitamin E (α-tocopherol)	25 mg
Distd. water	upto 250 ml

13. Glucose and phosphate buffered saline (PBS) solution:
First PBS solution is prepared by dissolving 1 tablet of PBB in 100 ml of distd. water

Glucose	18 mg
PBS solution	upto 10 ml

The final solution of glucose and PBS is sterilized by filtration through 0.22 mm disposable filters and stored at 4 ^{o}C.

14. Ampicillin and streptomycin solution (solution should be stored at -20 ^{o}C):

Ampicillin	500 mg
Streptomycin sulfate	250 mg

Distd. water upto 2.5 ml

15. Growth media is made by mixing the following solutions and adjusting the pH to 6.9 by adding 1M NaOH.

Basic TPS 1 medium (Diamond medium)	450 ml
Inactivated horse serum	50 ml
Vitamin mixture	15 ml
Ampicillin and streptomycin solution	1 ml

15. Centrifuge
16. Trypan blue dye
17. Incubator
18. Ice
19. Micropipettes (10, 50, 100 ml)
20. Microfilter (0.22 mm)
21. Saline solution
22. Liquid nitrogen
23. Biological freezer unit
24. Water bath
25. Test sample (crude extract, pure natural product or synthetic product)

The amoebicidal assay involves the following steps:

Maintenance of Culture

1. *E. histolytica* is cultured in Falcon flasks; it should be subcultured twice a week. The culture is examined daily through an inverted microscope in order to monitor the growth and detect the presence of contamination.
2. Subculturing is generally required when amoebae grow very rapidly and attach to the wall.
3. For subculturing the Falcon flasks are cooled in ice bath for 30 min. in order to detach amoebae from the walls of the flasks.
4. The suspension is centrifuged in sterile tubes at 1000 rpm for 5 min. The pellets of the amoebae are combined together and made up to 1 ml with medium and counted in a haemocytometer.
5. The volume of the suspension is adjusted with media to give a count of 5 amoebae per haemocytometer square.
6. 0.3 ml of this suspension is added to each Falcon flask containing culture media. The flasks are filled with media to exclude air and incubated at 37 ^{o}C.

Method for the Cryopreservation of Amoebae

1. The amoebae from one Falcon flask, when ready for subculturing, are centrifuged at 1000 rpm for 5 min. The pellet is made upto 1 ml with the medium.
2. 1 ml of sterile glucose and PBS solution containing 15% DMSO is added to the amoebal suspension.
3. This suspension is then placed in a sterile ampoule and thoroughly mixed. After staining at room temperature for 15 min, the suspension is frozen over liq. nitrogen to -70 ^{o}C by the use of a biological freezer unit.
4. The ampoules are left overnight and subsequently stored in liquid nitrogen.
5. Whenever required, an ampoule can be thawed by gentle shaking in a water bath at 37 ^{o}C, the content is transferred to a small Falcon flask (25 cm^2) which is then filled with medium and incubated at 37 ^{o}C. The fresh medium is given after 24 h and the amoebae subcultured as required.

Bioassay Procedure

1. Different concentrations of the test sample (0.01-100 mg/ml) are prepared in the appropriate volumes of growth medium with the aid of DMSO (40 ml). This solution is then sterilized by filtering through a 0.22 mm filter.
2. 0.1 ml of the amoebal suspension is added to each 12 ml flat-sided test tube (Nunc) which is filled with the medium to exclude the air.
3. These tubes are then incubated for 72 h at 37 ^{o}C with the flat surface down.
4. Each test is performed in duplicate and the control tubes do not have test samples.
5. After incubation, the amoebae are examined through an inverted microscope. The amoebae are detached from the tube walls by cooling in ice water for 30 min. and then centrifuged at 1000 rpm for 30 min.
6. The supernatant is removed carefully from each tube and the amoebal pellet suspended in 25 ml of the medium.
7. An equal volume of 1% Trypan Blue in normal saline is added and the amoebae counted on a haemocytometer.
8. The number of dead and alive amoebae are counted. Dead amoebae are stained blue and living amoebae can be easily

counted in each of the four corner squares of the haemocytometer slide. The average number of amoebae per square are calculated and divided by four to give the number ($x10^6$) in 1 ml of suspension.

9. Percent inhibition is calculated for each concentration.
10. The % inhibition value calculated for each concentration is transformed into probit values. The plot of probit (i) against log concentration is made and the best straight line determined by regression analysis. The regression coefficient, its level of significance (p) and the correlation coefficient are calculated and the 95% CI values determined.

1.6.6 Cercaricidal Assay

Another approach to control the tropical disease schistosomiasis is to interrupt the life cycle of the parasite *Schistosoma* by killing the cercariae. Since cercariae are generally short-lived, multiple treatment of the cercaricidal agent is necessary. Since the infected host molluscs continuously shed cercariae, it is often more effective to use an agent which can kill both cercariae and miracidiae (the free living transmission stages of the parasite). Natural products and extracts of some plants have been found to possess such activities (Squire *et al.*, 1989).

Materials

1. Mice (Swiss TO males), 6 weeks old
2. Circular plastic wells (1 ml)
3. Light microscope
4. Snails (*Biomphalaria glabrata*) 3-5 mm
5. Cold light, strong source
6. Artificial hard water medium
7. Lettuce
8. Rabbit pellets*
9. Test sample (crude extract, natural product or synthetic compound)

The steps which are involved in cercaricidal activity assay are following:

1. *Schistosoma mansoni* are maintained by passage through mice and snails (*B. glabrata*).
2. The mice are infected by paddling and sacrificed at 7 to 8 weeks after infection.
3. The eggs are recovered from the livers of sacrificed mice and hatched in fresh water.
4. Snails (3-5 mm) are separately exposed to 3-4 miracidia each, overnight.

5. The snails are maintained in artificial hard water medium at 27-28 °C and fed on lettuce and rabbit pellets.*
6. After 4 weeks the snails are checked for cercariae shedding by exposing the individual snails to a strong cold light source from 11.00 h onwards. The snails shedding cercariae are collected.
7. Known shedding snails are exposed to light for 15 min. and the cercariae released in this period are collected for use in the toxicity assay.
8. Various concentrations of test samples (100, 125, 200, 250 and 500 ppm) are prepared for testing. Upto 8 replicates are used for each concentration.
9. Each test is carried out in 1 ml plastic well. Approximately 20 freshly emitted cercariae are placed in each well and test samples added to a volume of 0.5 ml.
10. The control plastic well receives only freshly emitted cercariae.
11. Each cercariae is observed every minute (500 and 250 ppm conc.) or every 2 min. (200, 125 and 100 ppm conc.) at room temperature under a cold light source.
12. Observations continue until immobility is observed (the state when no swimming could be elicited by physical stimulation with a mounted eyelash).
13. Death is judged to have occurred when no muscle movement can be detected even after stimulation.

1.6.7 Leishmanicidic Activity

Leishmaniasis is a common disease endemic to the tropical areas of the Mediterranean region. Several species of the microorganism *Leishmania* such as *L. major, L. donovani* and *L. tropica* etc. are responsible of this disease. Only few drugs such as *N*-methylglucamine antimonate, amphotericin and pentamidine are clinically used to treat leshmaniasis. However, they are expensive and exhibit some toxicity to man and animals. The discovery of new antileishmanial medicines is therefore highly desirable. Leishmanicidic activity of natural products or crude extracts can be determined by the following bioassays:

1.6.7.1 In vitro Leishmanicidal assay (promastigotes)

Promastigotes are the extracellular flagellated form of the parasite and they can be used for *in vitro* determination of leishmanicidic activity of

---

* Rabbit pellets can be purchased from RGP Pellets, Grain Harvesters Ltd., Canterbury, UK.

natural compounds. *Leishmanial promastigotes* can be obtained from infected animals or human in endemic areas. They can also be obtained from *Leishmania* Reference Center at London School of Hygeine & Tropical Medicine, UK.

Materials

1. Tissue culture flasks 25 cm^2
2. 96 well microtiter plates
3. Medium M-199 buffered with 25 mM HEPES and 10% heat inactivated foetal bovine serum (HIFBS); pH 7.2
4. Double distilled deionized water
5. Dimethyl sulfoxide (DMSO)
6. Absolute methanol
7. Incubator
8. Light microscope
9. *Leishmanial promastigotes (*L. major, L. tropica, L. donovani, L. infantum, L. amazonesis, L. panamanensis*)
10. Amphotericin B, pentamidine
11. Test sample (plant extract, pure natural product or synthetic compounds)

The following procedure is used for the leishmanicidal assays:

1. Leishmanial promastigotes are cultured in sterile 25 cm^2 tissue culture flask in tissue culture medium M-199 supplemented with 25 mM HEPES and 10% HIFBS at 25°C.
2. Parasites are centriguged at 3000 rpm, diluted in minimum volume of PBS and are counted with the help of improved Neubauer chamber under a microscope.
3. Parasites are diluted with the fresh medium to a final concentration of 2.0 x 10^6 parasites ml^{-1}.
4. 1.0 mg of compound is dissolved in 50 μl of absolute MeOH or DMSO and the volume is made upto 1.0 ml with the culture medium.
5. In a 96 well microtiter plate, 90 ul of the parasite culture (2.0 x 10^6 parasites ml^{-1}) is placed and 10 μl containing various concentrations of the experimental compound is added in the culture. 10 ul of PBS (phosphate buffered saline, pH 7.2 containing 0.5% MeOH/ 0.5% DMSO) is added as negative control while amphotericin B, and pentamidine (to a final

concentration of 1.0 mg ml^{-1}) are added separately as positive control.

6. The plates are incubated at 25°C in the dark for 3-5 days during which control organisms multiply 3-6 times. The culture is examined microscopically on an improved Neubauer chamber and ED_{50} value of compounds possessing antileishmanial activity is calculated.

1.6.7.2 In vitro Leishmanicidal assay (amastigotes)

Mononuclear phogocytes from human peripheral blood and rodent peritoneal macrophages are frequently used for short term assays for the growth of amastigotes. A great variety of animal cell lines have also been tested for their capacity to grow leishmaniae. Cells of the J774G8 lines are particularly easy to handle.

Materials

1. Tissue culture flasks 25 cm^2
2. 96 well microtiter plates
3. Medium RPMI-1640 buffered with 2.2 g l^{-1} $NaHCO_3$ and 10% heat inactivated foetal bovine serum (HIFBS); pH 7.4
4. Double distilled deionized water
5. Dimethyl sulfoxide (DMSO)
6. Absolute methanol
7. Giemsa stain
8. CO2 incubator
9. Light microscope
10. *Leishmanial promastigotes (*L. major, L. tropica, L. donovani, L. infantum. L. amazonensis, L. panamanensis*).
11. J774G8 cell line
12. Amphotericin B, pentamidine
13. Test sample (plant extract, pure natural product or synthetic compound).

Another procedure for leishmanicidal assay is as follows:

Cell Culture

A. *Human blood derived mononuclear macrophages*

1. Blood is drawn from a normal human in the presence of 1.0 mg ml^{-1} EDTA and centrifuged at 800 rpm for 5 minutes. The plasma

is collected and diluted 1:2 with sterile phosphate buffered saline (pH 7.4). Diluted plasma is loaded over Ficoll-Paque (1:3 v/v) and centrifuged at 1000 rpm for 30 minutes at room temperature in a centrifuge with brakes off. Mononuclear lymphocytes and monocytes are aspirated at the Ficoll-Plasma interface where a white band is formed with a sterile Pasteur pipette. The cells are washed with sterile normal saline at 2000-2500 rpm for 10 minutes and suspended in RPMI-1640 medium. Cells are washed with RPMI-1640 twice and suspended in RPMI-1640 supplemented with 10% heat inactivated foetal bovine serum. The cells are cultured on plastic wells (1.5 x 10^6 cells/well) for 6 days. During this time half of the monocytes are adhere to the bottom and have enlarge into macrophages, the rest are removed by washing. The 6 day old macrophages are then infected wth *Leishmania* which multiply 1.5-3.0 times during the subsequent 6 days.

B. *J774G8 cell lines*

1. Culture flasks with approximately 75% confluence is taken and old culture medium is removed. The cells are washed twice very gently with PBS (Ca/Mg free; pH 7.4) and resuspended in fresh culture medium (RPMI-1640 + 10% HIFBS) by aspiration. The number of cells are counted on an improved neubauer chamber and divided 1:2 or 1:4 (depending on the culture volume) with fresh culture medium. The culture is placed at 37°C in a humidified CO_2 (5%) incubator.

Infection with leishmania

1. 10^6 macrophages (human blood derived/ J774G8) are taken and mixed with 10^7 leishmanial promastigotes. The culture is left for three days at 37°C in a humidified CO_2 (5%) incubator during which promastigotes find their entry in the macrophages and transformed into intercellular amastigotes. Within three days almost 80% promastigotes parasitize cells and 5-6 amastigotes/cell could be seen. The medium is changed after every three days.

96 Well Plate Assay

1. Compounds to be checked are dissolved to a final concentration of 1.0 mg in 1 ml of PBS (phosphate buffered saline, pH 7.4 containing 0.5% MeOH or 0.5% DMSO).

2. In a 96 well microtiter plate, 100 µl of infected macrophage culture (1.5 x 10^6 cells/well) are added in different wells. 10 µl portions containing various concentrations of the experimental compouned are added in the culture. 10 µl of PBS (phosphate buffered saline, pH 7.4 containing 0.5% MeOH/0.5% DMSO) is added as negative control while glucantime, amphotericin B, pentamidine or paromomycin sulfate to a final concentration of 1.0 mg ml^{-1} is added separately as positive control.
3. The plates are incubated between 36-37°C in humidified chamber in presence of 5% CO_2 for 6 days during which control organisms multiply 1.5-3.0 times. The culture is examined microscopically after staining with Giemsa and ED_{50} value of compounds possessing antileishmanial activity is calculated.

1.6.7.3 Leishmanicidal activity in vivo

Of all the laboratory animals e.g. monkey, rabbit, guinea pig, rat, mice and hamster etc., inbred BALB/c mice and golden hamsters are the favorite animal models for researchers working on leishmaniasis. What makes these animals most suitable are their manifold genetic varitions, convenient size, high fertility rate, short gestation period, easy maintenance, resistance to many infections and representation of diseases similar to humans and, above all, their selective susceptibility.

Materials
1. Balb/c mice
2. Golden hamster
3. Dimethyl sulfoxide (DMSO)
4. Giemsa stain
5. Light microscope
6. *Leishmanial promastigotes (*L. major, L. tropica, L. donovani, L. infantum. L. amazonensis, L. panamanensis*)
7. Test sample (plant, pure natural product or synthetic compounds)

The procedure for antileishmanicidal activity (*in vivo*) is given below:

1. 100 ul inoculum containing 5.0 x 10^6 log phase parasites ml^{-1} are administered intradermally on the front paw or on the nose of the golden hamster or on the root/base of the tail (cutaneous models), or intracardially or intravenously (visceral models) or intranasally (mucocutaneous models) of BALB/c or NMRI inbred mice.

2. The lesion is allowed to develop fully for 2-4 weeks.
3. Beginning either a few days or few weeks after infection, the antileishmanial agent or compound to be tested is administered either subceuaneously or as per model.
4. The effect of the drug is monitored by observing the decrease in lesion size in experimental animals compared to untreated controls (placebo) or by monitoring microscopically the number of infected macrophages obtained from the lesion or by culturing the inoculum removed from the lesion [Fournet, *et al.*, 1994].

1.7 ASSAYS FOR AGROCHEMICALS

1.7.1 *Lemna minor* for Phytotoxicity and Growth Stimulating Assay

Weeds are one of the major factors of poor agricultural productivity in the developing world. Synthetic weedicides (herbicides) are often expensive, toxic and non-specific. Weedicides from natural sources having improved characteristics could therefore have a promising future.

The search for new weedicides/herbicides requires an assay which can predict the general phytotoxic effects of the test samples. The specificity for certain weeds can be determined at later stages. The *Lemna minor* phytotoxicity assay is a useful primary screen for weedicide search. This bioassay has the added advantage of being able to predict the growth stimulating effect of the test sample.

The *Lemna* assay is a quick measure of phytotoxicity of the materials under investigation. The assay can be used for testing column fractions or pure compounds in concentrations of 1 to 1000 ppm. The assay is run for 7 days and the rosset or frond number is counted on the third and seventh days. For consistency a rosette of three fronds per replicate should be used. For all values on frond number it is best to count all fronds, no matter how small, in order to avoid subjective analysis. The assay has the advantage that very small sample quantities are required for testing.

Lemna minor L. (duckweed) (Lemnaceae) is a miniature (1.5x1.5 mm) aquatic thalloid monocot. *Lemna* plants consist of a central oral frond or mother frond with two attached daughter fronds and a filamentous root. These plants are generally found in water ponds and other fresh water bodies (Einhelling, 1985).

Materials

1. *Lemna minor* L*

2. E-Medium **(about 80 ml per compound)
3. Syringes: 10 ml , 100 μl, 500 μl, 2 μl
4. 2 dram vials (40 per compound)
5. Large glass container with glass lid to hold vials; (stopcock grease should be applied on the top edges of the tank in order to form a seal with the lid to avoid moisture loss).
6. Growth chamber with temperature range of 27° to 29°C and continuous fluorescent and incandescent light
7. Test sample (crude extract, pure natural product or synthetic compound)

The commonly used procedure for *Lemna assay* is as follows:

1. Prepare inorganic medium (E- Medium)*: add KOH pellets to attain pH 5.5 - 6.0.
2. Prepare vials for testing: 10 vials per dose (500, 50, 5 ppm, control)#.
   a. Weigh 15 mg of compound or extract and dissolve in 15 ml solvent.
   b. Add 1000, 100, and 10 μl solutions to vials for 500, 40 and 5 ppm; allow solvent to evaporate overnight.
   c. Add 2 ml of E- medium and then a single plant containing a rosette of three fronds to each vial. Only healthy and green rosettes should be used.
3. Place vials in a glass dish filled with about 2 cm water, seal container with stopcock grease and glass plate.
4. Place dish with vials in growth chamber for seven days at 26°C under fluorescent and incandescent light.
5. Count and record number of fronds per vial on day 3 and day 7.
6. Analyze data as percent of control with ED_{50} computer program† to determine FI_{50} values and 65% confidence intervals.

#500 ppm = 15 mg test sample/15 ml solvent --->1000 μl of this solution/2 ml E medium; 50 ppm = 100 μl of this solution/2 ml E medium; 5 ppm = 10 μl of this solution/2 ml E medium; control vials contain no test sample.

†A copy of this program for IBM PC's is available from Prof. Jerry L. Mclaughlin, Department of Medicinal Chemistry and Pharmacognosy, West Lafayette, Purdue University, IN 47907, U.S.A.

***E- MEDIUM**

Constituent	mg/l
KH_2PO_4	680
KNO_3	1515
$Ca(NO_3)_2 .4 H_2O$	1180
$MgSO_4 .7 H_2O$	492
H_3BO_3	2.86
$MnCl_2$	3.62
$FeCl_3 .6 H_2O$	5.40
$ZnSO_4 . 7 H_2O$	0.22
$CuSO_4 . 5 H_2O$	0.08
$Na_2MoO_4 .2 H_2O$	0.12
EDTA	11.2

1.7.2 Contact Toxicity Insecticidal Assay

This simple test is used to assess the direct insecticidal actions of pure natural products or plant extracts. This method unambiguously demonstrates if a compound or extract is lethal to certain types of insects *on contact* and not because of the volatility of the samples (Isman *et al.*, 1987).

Materials

1. Volatile organic solvents (diethyl ether, acetone, methylene chloride, methanol, etc.).
2. Glass vials (20 ml capacity)
3. Test insects
4. Test sample (crude extract, natural product or synthetic compound)

The contact toxicity assay is performed according to the following:

1. Suitable quantities (1000, 500, 100, 50, 10 ppm) of test sample are dissolved in a volatile organic solvent and these solutions are then coated on the inner surface of 20 ml glass vials (two to five replicates for each concentration).
2. Each glass vial is rotated by hand until the test solution is distributed on the vial innerwall and floor, and the solvent has mostly evaporated. Then each vial is placed in a fume hood for 10 min. to ensure complete removal of the carrier solvent. Care should be taken to ensure that a uniform quantity of test material is coated on the inner surface of the vial.
3. When the solvent has completely evaporated, five test insects (or larvae) are placed carefully in each vial with sufficient food (i.e.

the natural diet for that particular insect, such as leaves, grain, etc., or artificial diet which is different for different insects).
4. The survival of the insects is assessed after 24-48 hours.
5. Controls consist of test insects (or larvae) in vials, treated only with the carrier solvent. Survival of such controls should average over 95%.
6. LC_{50} values can be calculated by using a Finny computer program mentioned in the case of brine-shrimp lethality assay (Section 1.1.1).

1.7.3 Insect Antifeedant Assays

1.7.3.1 Leaf-choice assay

Antifeedants are chemicals which when combined with a favorite food, renders the food undesirable to insects. Such chemicals can therefore be used for pest management. There are several methods to test for antifeedant activity. A simple leaf choice assay is appropriate to isolate antifeedant compounds from natural sources (Kubo, 1991)

Materials
1. Incubator (with humidity control).
2. Homogenizer
3. Cork borer (1 cm)
4. Petri dishes
5. Volatile organic solvent such as acetone, ether, etc.
6. Filter paper
7. Host specific insects, for example *Pectinophora gossypiella* (Sunders), family Gelechiidae, for cotton (*Gossypium baradanse*)#
8. Uncontaminated, disease-free fresh leaves of the host plant, such as cotton, generally from a green house
9. Test sample (crude extract, pure natural product synthetic compound)

The commonly used host-guest test systems are the larvae of African armyworms (*Spodoptora littoralis* and *S. exepta*) on leaves of *Ricinus communis* and *Zea mays*, the Mexican bean beetle (*Epilachna varivestis*) on leaves of stringbeans, the silk bollworm (*Pectinophora gossypiella*) and the tobacco budworm *Hellothis virescens* on cotton leaves.

The following steps are involved in the leaf-choice assay:

1. Discs (1 cm) are punched out by a cork borer from the leaves of the host plants. The leaf discs should be similar in size, shape, thickness and chlorophyll content.
2. The leaf discs are randomized and arranged (12 discs per dish) in a circle on a moistened filter paper in polyethylene which forms a grid inside the glass petri dishes (100 x15 mm).
3. Alternating discs are treated on their upper surface with either acetone (25 ml) or with 1-100 μg of a test sample dissolved in 25 ml of acetone, ether or any other suitable volatile organic solvent (applied with a microsyringe).
4. Newly-moulted third-instar larvae, reared from eggs on a meridic artificial diet*, are placed, three larvae per dish, at 25 °C and 80% relative humidity in a dark incubator.
5. After 48 hr, the larvae are removed and the discs are examined visually.
6. Activities of the antifeedant are compared in terms of 95% protective concentration (PC_{95}) values. Experiments determining PC_{95} values should be repeated at least six times.

1.7.3.2 Artificial diet feeding bioassay

The specific diet of each insect varies, and it is sometimes difficult to find the right type of diet (host plant or crop) for an insect to carry out the leaf-choice bioassay. It is therefore desirable to prepare an artificial diet on which most of the common insects can be reared and subsequently used in testing the antifeedant activity of plant extracts and other natural products (Kubo, 1991; Chan *et al.*, 1978).

Materials

1. The eggs of test insects such as *Pectinophora gossypiella* (Saunders) family Gelechiidae (serious pest of cotton), *Heliothis virescens* (Fabr.) family Noctuidae (common crop pest), *Bombyx ori* Linnaeus (common silk worm) or any other insects of agricultural importance#,+

---

*The procedure is given in the next section.

#The eggs can be stored at 10° -15 °C for several days without any effect.

+The insect eggs can be obtained from the Entomology Departments of different universities or from the Plant Protection Departments of Agricultural Research Councils. Requests for eggs can also be sent to the ARS, Brownsville, TX, USA and ARS, Phoenix, AZ, USA.

2. α-Cellulose
3. Organic solvents (such as methanol, diethyl ether, ethyl acetate, ethanol etc.)
4. Beakers
5. Desiccator
6. Gelling agent: Agar 4 gm/100 ml the suspension is autoclaved for 15 min and kept at 70 °C on a steam bath until used.
7. Freshly prepared vitamin solution (final volume 100 ml) containing:
   Ascorbic acid 1.76 g/95 ml distd. water, neutralized to pH 6.5 by adding 1 N KOH

Chlorotetracycline	20 mg
30% Choline chloride solution	1 ml
Conc. vitamin suspension	0.5 ml

8. Conc. vitamin suspension (in 1 ml of dist. water)

Niacin	6 mg
Calcium pantothenate	6 mg
Folic acid	1.5 mg
Riboflavin	3 mg
Thiamine hydrochloride	1.5 mg
Pyridoxine hydrochloride	1.5 mg
Biotin	0.12 mg
Vitamin B_{12}	0.012 mg

9. Solid nutrients: The following ingredients are mixed, blended and pulverized to a fine powder. Generally 1 kg is prepared at a time. Store at -20 °C.

Wesson's salt mixture	100 g
Wheat germ	300 g
Sucrose	350 g
Vitamin-free casein	350 g
Potassium sorbate	14 g
Methyl *p*-hydroxybenzoate	15 g

10. Glass rod
11. Incubator or a pressure cooker
12 Autoclave
13. Stereo-dissecting microscope (magnification 10-30x)
14. Refrigerator
15. Soft hair brush (fine camel-hair brush)
16. Plastic catsup cup
17. Dissecting needle
18. Test sample (crude extract, pure natural product or synthetic compound)

The systemic protocol adopted for artificial diet feeding bioassay is as follows:

1. The test sample is dissolved in a small beaker in a minimum amount of methanol (or any appropriate solvent) in order to prepare sample solution.
2. This sample solution is added to a 125 ml beaker in which α-cellulose is previously weighed (1 g). The organic solvent is then removed under reduced pressure using a desiccator. A separate beaker is required for each test sample and one beaker is required for the control.
3. The solid nutrient (11.7 g) and vitamin solution (33 ml) are added to the beaker. Finally 50 ml gelling agent (4% agar) is added and quickly mixed by vigorous stirring with a glass rod. The slurry in the beaker should form a homogenous jelly. A layered diet is not acceptable. Warming and cooling can be used to control the gel formation.
4. The diet jelly is cut into 10 equal pieces by a sterile knife. The pieces of diet are then transferred to appropriate containers (2-dram disposable plastic scintillation vials) with tweezers or a dissecting needle. Two larvae of each species are used per cup.
5. For inoculation, one-day old insect larvae* are carefully transferred with a fine camel-hair brush to vials containing excess portion of the diet.
6. After larval transfer the caps are pierced with a dissecting needle for ventilation. The smaller of the two larvae is removed after 3-4 days to reduce variability. Each test sample needs ten replicates.
7. The larvae are incubated for 8-10 days at 28 ^{o}C in an incubator under a12-14 h day length and the relative humidity is maintained at 90-95%.
8. The larval weights are determined during the assay period with a balance and daily observations are made with a stereo-dissecting microscope.
9. The effective doses of the isolated test sample that cause a 50% reduction in insect growth (ED_{50}) need to be determined. The weights of 20 larvae in mg after 14 days at 26°C (mean ± SD) are plotted against the % of the test sample in the diet and the ED_{50} values are determined from a curve of the log-dose against weight of insect lines. Usually at least three doses causing reduction of weights between 20% and 80% are used, with three replicates of 20 insect larvae at each dose. The lethal doses causing 90% death (LD_{90}) are determined if larval mortality is high.

---

*One-day old tiny larvae are very sensitive to the effects of handling.

1.7.4 Nematicidal Assay

Nematodes are the most numerous multicellular animals on earth. A handful of soil will contain thousands of the microscopic warms (nematodes), many of them parasites of insects, plants or animals.

Root-knot nematodes are plant parasites which cause heavy damage to crops and vegetable and fruit plants world wide. Currently used synthetic nematicides are generally harmful to human health and causing environmental problems. There is therefore a need to develop nematicides from natural sources. Following assays provide preliminary indication of nematicidal activity of crude extracts and are easy to perform.

Any of the root-knot nematode species such as *Meloidogyne incognita* or *M. javanica*, which produce root-knot galls on brinjal (*Solanum melongena*) plants can be used to test the nematicidal properties of test samples (Taylor *et al.*, 1978).

Materials

1. Root-knot nematode culture
2. Sodium hypochlorite
3. Cavity blocks
4. Seeds of test plants such as brinjal, okra, chilli, tomato, etc.
5. Brinjal plants infested with root-knot nematodes
6. Distilled water
7. Beaker (200 ml)
8. Wide mouth bottle
9. Organic solvents such as methanol, chloroform, etc.
10. Compound microscope
11. Test sample (crude extract, pure natural product or synthetic compound)

The following method is used for the nematicidal assay:

a. *Preparation of egg suspension*

1. Roots of brinjal plant infested with *M. javanica* showing symptoms of root-knot galls are cut into small pieces.
2. These pieces are placed in a wide mouth bottle containing 1% sodium hypochlorite solution and the bottle mouth tightly closed.
3. The bottle is shaken vigorously for 3 min. and the contents poured onto a 100 mesh sieve fitted over a 400 mesh sieve.
4. The eggs collected on the 400 mesh sieve are washed under running tap water for 1 min. and then transferrd to a beaker for use.

5. The number of eggs per ml suspension is determined with the help of a counting chamber.

b. *In vitro Nematicidal Activity Test*

1. Different concentrations of test sample in organic solvent are poured into cavity blocks (2 ml in each).
2. The cavity blocks are kept open overnight to allow the organic solvent to evaporate.
3. Two ml of egg suspension of root-knot nematode is poured into cavity blocks and the hatchability of eggs examined under a microscope at 24 hr. intervals upto a period of 6-7 days.
4. The mean effective concentration (MEC) of the test sample at which the first effect appears is calculated.
5. The egg suspension is kept for 2-3 days to obtain hatched 2nd stage (J_2) larval suspension of the juveniles.
6. Two ml of the larval suspension is poured into the cavity blocks and the mortality of J_2 larvae examined under the microscope at 24 hr. intervals over a period of 6-7 days.

c. *In vivo Nematicidal Activity Test*

1. Test plants such as brinjal, chilli, mungbean, okra or tomatoe are grown in small pots.
2. One week old seedlings are inoculated with root-knot nematode suspension containing 100 eggs or juveniles per pot followed by treating with test sample.
3. Plants not inoculated with nematode and not treated with test sample are kept as control.
4. After 8 weeks plants are uprooted and root-knot index (RKI), length and fresh weight of plants are determined.

d. *Calculation of Root-Knot Index (RKI)*

Infestation of roots by root-knot nematode is estimated on a 0-5 scale as described by Taylor, A. L. and Sasser, J. N. (1978).

e. *Effect of Test Samples on Nematode Population in situ*

1. A 500 gm ml^{-1} soil sample is collected from the field. The soil is stirred in a bucket half filled with water and then poured onto a 100 mesh sieve fitted over a 400 mesh sieve and washed several times under running tap water.

Table 1. Root-knot Index

Number of knots per plant	RKI
0	0
1-2	1
3-10	2
11-30	3
31-100	4
> 100	5

2. The residue collected on a 400 mesh sieve is transferred onto a Buchner funnel and allowed to stand.
3. After about 48 hours, 10 ml of water is collected in a beaker and the number of nematodes determined with the help of a counting chamber.

1.8 HEPATOTOXICITY ASSAYS

1.8.1 Antihepatotoxic Activity Using Carbon Tetrachloride-Induced Cytotoxicity

A number of plant extracts and other natural substances have hepatoprotective activity. Some crude drugs used in the indigenous systems of medicine are also believed to be effective against hepatitis. Various assays have been devised to examine crude drugs and pure natural products for hepatoprotective activity. Carbon tetrachloride (CCl_4) is a strongly hepatotoxic agent and it is therefore used to induce cytotoxicity in primary cultured hepatocytes in order to monitor liver protective effects of various natural substances (Kiso *et al.*, 1983).

Materials
1. Mice (20-25 g, ddY)
2. Wistar Rats (200-250 g)
3. Bovine serum albumin Fr. V
4. Calf serum
5. Carbon tetrachloride (CCl_4)
6. Collagenase type I (125-250 IU/mg,)
7. Dexamethasone (Sigma)
8. Eagle's minimum essential medium (Eagle's MEM, Nissui)
9. Ethylene glycol-*bis*-(β-amino ethylether)*N*,*N'*-tetraacetic acid (EGTA), (Dotite)

10. Phase contrast microphage
11. Insulin (Novo Research)
12. Potassium penicillin G
13. Streptomycin sulfate
14. Ethanol
15. DMSO
16. Plastic dishes (Falcon)
17. Humidified incubator
18. Trypan blue
19. Hank's buffer (Ca^+ free)
20. Autoanalyser
21. Teflon catheter
22. Forceps
23. Rotator
24. Ether
25. Test sample (crude extract, pure natural product or synthetic compound)

The following steps are involved in the antihepatotoxic activity assays:

Method of Isolation and Culturing of Hepatocytes

1. The abdomen of the mouse or rat is opened under ether anesthesia. A midline incision is made and the portal vein is cannulated with a needle fitted with a teflon catheter.
2. The teflon catheter is tied in place and the needle is removed. The inferior vena cava is cut below the renal vein.
3. The liver is perfused using Ca^{2+}- free Hank's buffer containing 1% bovine serum albumin Fr. V and 0.5 mM EGTA. Aeration is carried out with 95% O_2/5% CO_2 at a flow rates of 30 ml/min. to pH 7.4 at 37 °C.
4. The thorix portion of the superior vena cava is cannulated, and the inferior vena cava is tied off above the renal vein.
5. After perfusion of the liver for 10 min., the Ca^{2+}- free Hank's buffer (100 ml) (containing additionally 0.075% collagenase and 4 mM $CaCl_2$) is recirculated.
6. After 10-15 min. perfusion, the liver is transferred to a beaker containing Ca^{2+}- free Hank's buffer (50 ml) and gently dispersed with two forceps.
7. The crude cell suspensions are then rotated in a rotator under oxygen-carbon dioxide at 37 °C for 10 min.

8. The cell suspension is then cooled in ice and filtered gently through cotton gauze into centrifuge tubes.
9. The preparation is centrifuged at 50 g for 1 min. The supernatant is removed and the loosely packed pellet of cells is gently resuspended in Ca^{2+}- free Hank's buffer.
10. The washing procedure is repeated 3 to 5 times.
11. The viability of cells to exclude trypan blue is determined by incubating the cell suspension (0.1 ml) with 0.4% trypan blue (0.9 ml) and then counting those cells the nuclei of which were stained as well as those which excluded the dye.
12. 1-30 mM CCl_4/ethanol (concentration 1%) is dissolved in a culture medium# (1.0 ml) containing DMSO (0.01 ml).
13. The cells are preincubated (1.5 or 24 hr.) and then exposed to the above medium containing CCl_4 and the test sample.
14. The glutamic-pyruvic transaminase (GPT) activity in the medium is measured after every 15 min over a 90 min. period using the method of Karmen, A. F. *et al* (1955), employing an autoanalyser, and the liver cells examined using a phase contrast microscope.
15. The GPT values (IU/l x 10^{-2}) are plotted against the dose administered and against time to determine the dose and time dependence of CCl_4-induced cytotoxicity in primary culture hepatocytes.

1.8.2 *In Vivo* Hepatoprotective Assay

The following assay can be used to evaluate the efficacy of natural products or crude extracts in preventing damage induced by CCl_4 with the help of morphological, biochemical and functional parameters (Rana *et al.*, 1992).

Materials
1. Albino rats of either sex (60-80 g)
2. Liquid paraffin
3. Carboxymethylcellulose (CMC)
4. Syringes with needles

The culture medium is composed of Eagle's MEM supplement with 10% calf syrup (which has been inactivated by warming to 56 ^{o}C for 30 min.), penicillin (100 IU/ml), streptomycin (100 μg/ml), 10^{-6} M dextramethasone and 10^{-8} M insulin. Inocula of 5×10^4 cells 0.1 ml/cm^2 are sealed into plastic dishes (Falcon Plastics) and preincubated in a humidified incubator at 36 ^{o}C under 5% CO_2 in air for 1.5 or 24 hr. (when preincubated for 24 hr., the medium is replaced after 6 hr.) (1.0 ml).

5. Distilled water
6. Carbon tetrachloride (CCl_4)
7. Pentobarbitone sodium
8. Weighing balance
9. Heparin sodium
10. Test sample (plant extract, pure natural product or synthetic compound)

The steps which are involved in the hepatoprotective activity assay are the following:

1. The rats are divided into three groups, each group having six animals:

Group A	=	Normal animals
Group B	=	CCl_4 intoxicated control animals
Group C	=	Plant extract (Test sample) treated animals

2. Liver damage or intoxication in group B is produced by injecting CCl_4 (1 ml/kg with equal volume of liquid paraffin) twice weekly for a period of eight weeks. Group B receives the vehicle (CMC suspension) and sub-cutaneous injection of CCl_4.
3. A 5% suspension of test sample is prepared with 1% CMC in distilled water.
4. The animals of group "C" are treated twice weekly with test sample suspensions (300 mg/kg) for a period of eight weeks. The animals also simultaneously receive CCl_4 suspensions twice weekly for eight weeks.
5. All the animals are injected with 100 units of heparin sodium. After 30 minutes the animals are anaesthetized with ether. 4 ml portions of the blood are collected with a syringe in heparinised tubes by direct heart puncture. The samples are immediately used for biochemical estimations.
6. The livers of each group of animals are removed and observed carefully for any change in appearance e.g. necrosis and weight. Their volumes are measured by the water displacement method.
7. CCl_4 damage and efficacy of test sample are evaluated with the help of the following three parameters:

Morphological Parameters: Weights of animals, and weights and volumes of livers of scarified animals after eight weeks are recorded. If the test sample is hepatoprotective, the livers of group C animals will have no visible signs of toxicity such as necrosis. Similarly the percent change in the group A and C weights of the animals, as well as their liver weights and liver volumes should be insignificant.

Functional Parameters: Pentobarbitone sleeping time test is performed. This test is used to assess the efficacy of the functional liver cell. Pentobarbitone sodium (50 mg/kg) is given to the animals of each group interperitoneally. The sleeping time (time interval between loss and gain of writhing reflex) is determined.

If the test sample is hepatoprotective the sleeping time (min.) of animals of groups A and C should be the same, while for the CCl_4 intoxicated Group B it will be significantly longer. The increase in sleeping time is a sign of liver damage.

Biochemical Parameters: The serum enzyme level of alanine transferase (SGPT) and alkaline phosphatase (AP) should be measured# . The increase in SGPT and AP levels in the CCl_4 control group indicates the extent of liver damage. If the test sample is hepatoprotective the SGPT and AP activity levels (after eight weeks) in groups A and C would be almost the same, while the SGPT and AP levels in the CCl_4 control group B is raised significantly.

1.8.3 Antihepatotoxic Activity Using Galactosamine Induced Cytotoxicity

This *in vitro* assay method is used for the primary screening of antihepatotoxic activity of extracts using D-galactosamine-produced injury in primary-cultured mouse and rat liver cells. This bioassay model system is now recognized to resemble the effects of viral hepatitis in humans from both the morphological and functional points of view. It is therefore considered to be a very useful bioassay screen for assessing prevention of liver damage and is comparable with the *in vivo* assay method.

Materials

1. Male Std: dd Y mice (20-25 g)/Male Std: Wistar rats (200-250 g)
2. Bovine serum albumin Fr. V 1%
3. Calf serum 10% (Flow Lab.)
4. Collagenase type I (125-250 IU/mg) 0.075%
5 Dexamethasone 10^{-6}M
6. Ethylene glycol-*bis*- (β-amino ethylether) *N, N'*-tetraacetic acid (EGTA) 0.5 mM

Check literature for specific method.

7. Insulin 10^{-8}M (Novo Research)
8. Potassium penicillin G (100 IU/ml)
9. Streptomycin sulfate (100 μg/ml)
10. D-Galactosamine 0.1-0.5 mM (GalN)
11. $CaCl_2$ 4mM
12. Trypan blue 0.4 %
13. Ca^{2+}-*free Hank's buffer*
    Contains 1% bovine serum albumin Fr.V and 0.5 mM EGTA and is aerated with 95% O_2/5% CO_2 to pH 4.7 at 37°.
14. *Eagle's minimum essential medium* (Eagle's MEM) (Nissui)
    Eagle's MEM is supplemented with 10% inactivated calf serum (56° for 30 min), penicillin (100 IU/ml), streptomycin (100 μg/ml), 10^{-6} M dexamethasone, and 10^{-8} M insulin.
15. Liver protective natural products (LPNP) such as:
    a. Cynarin
    b. Glycyrrhetinic acid
    c. Glycyrrhizin
    d. Methionine
    e. Picroside I
    f. Picroside II
    g. Silybin
    h. Desoxypodophyllotoxin (isolated from *Podopyllum hexandrum*, E. Merck)
16. Equipment
    a. Needle fitted with a teflon catheter
    b. Scissors, forceps
    c. Beakers
    d. Rotator
    e. Cotton gauze
    f. Centrifuge tubes
    g. Centrifuge (upto 30 g values)
    h. Humidified CO_2 incubator
    i. Plastic dishes
    j. Auto analyzer (RaBA super)
17. Test sample (crude extract, pure natural product or synthetic compound).

This assay is carried out in the following three steps (A-H):

A. Isolation of Hepatocytes

Liver cells are isolated by a modified procedure of Seglen (1976).

1. The abdomen of the mouse or rat is opened under ether anaesthesia.
2. A midline incision is made, and the portal vein is cannulated with a needle fitted with a teflon catheter.
3. After the teflon catheter is tied in place and the needle removed, the inferior vena cava is cut below the renal vein.
4. Perfusion of the liver is started with Ca^{2+}- free Hank's buffer (Hanks *et al.*, 1949). The flow rate is 30 ml/min.
5. The thoracic portion of the superior vena cava is cannulated, and the interior vena cava is tied above the renal vein.
6. After the liver has been perfused for 10 min, recirculation of Ca^{2+}- free Hank's buffer (100 ml), which also contains 0.075% collagenase and 4 mM $CaCl_2$, is started.

Please note:
In the case of mice, the same buffer is poured into the inferior vena cava and drained from the hepatic portal vein at the flow rate of 10 ml/min.

7. After 10-15 min of perfusion, the liver is transferred into a beaker containing Ca^{2+}- free Hank's buffer (50 ml) and gently dispersed with two forceps.
8. Next, the crude liver cell suspension is rotated on a rotator under oxygen-carbon dioxide at 37°C for 10 min.
9. The cell suspension is then cooled on ice and gently filtered through cotton gauze into centrifuge tubes.
10. The preparation is centrifuged at 50 g for 1 min. (for mice, 30 g, 1 min).
11. The supernatant is aspirated off and the loosely packed pellet of cells is gently resuspended in Ca^{2+}- free Hank's buffer.
12. This washing procedure is repeated three to five times.

B. Determination of Viability of Isolated Cells

Viability of cells is determined by incubating cells suspension (0.1 ml) with 0.4% trypan blue (0.9 ml) and then counting the number of cells that excluded the dye and the number of cells whose nuclei are stained blue.

C. Preparation of Culture of Hepatocytes

The culture medium is composed of Eagle's MEM. Inocula of $5x10^4$ cells/0.1 ml/cm^2 is seeded into plastic dishes and preincubated in a humidified incubator at 36.5° under 5% CO_2 in air for 1.5 or 24 h. When preincubated for 24 h, the medium is replaced after 6 h.

D. Determination of Viability of Cultured Cells

Viability of cultured cells is assayed by trypan blue exclusion. After the direct addition of 0.4% trypan blue, the number of cells, stained and unstained is counted.

E. Determination of Cytotoxicity Induced by Galactosamine (GALN)

1. *Preparation of GalN Containing Media*

   In the medium (1.0 ml), a solution of 0.1-0.5 mM GalN dissolved in DMSO (0.01 ml) is added.
2. After preincubation (1.5 or 24 hr), the cells are exposed to the above-mentioned medium containing GalN.
3. At the indicated times after the GalN challenge, glutamic-pyruvic transaminase (GPT) activity in the medium is measured by the method of (Karmen *et al.,* 1955) using an autoanalyzer (RaBA Super, Chugai Pharmaceutical).

F. Determination of the Cytotoxicity Induced by Liver Protective Natural Products (LPNP)

As the reference, a similar experiment was conducted to determine the antihepatotoxic actions of the natural products known to exert liver-protective effects *in vivo*, utilizing GalN- intoxicated hepatocytes *in vitro.*

G. Determination of the Cytotoxicity Induced by Test Sample

A similar experiment is conducted, except that the test sample is used in place of GalN to examine its hepatotoxic action on hepatocytes and the GPT activity is measured.

H. Determination of Hepatoprotectivity Induced by the Test Sample

After preincubation of isolated rat hepatocytes for 1.5 h, the test sample is added at doses of 0.01, 0.1 and 1.0 mg, with or without 0.5 mM GalN, to the culture medium (1.0 ml). GPT activity is measured at 30 h after the treatment. If the test sample is hepatoprotective the GPT activity levels with or without GalN would be almost the same.*

1.9 HYPOGLYCEMIC / ANTIDIABETIC ACTIVITY ASSAYS

Since ancient times, patients with non-insulin requiring diabetes have been treated orally in folk medicine with a variety of plant extracts. The evaluation of these plants and especially of their active natural principles is a logical way of searching for new drugs to treat this disease. A number of bioassays have been established for this purpose (Akhtar *et al.*, 1981).

1.9.1 Antidiabetic Activity Assay on Normal and Alloxan-Diabetic Rabbits

Materials

1. Alloxan-monohydrate (BDH)
2. Carboxymethylcellulose (CMC)
3. α-D-Glucose
4. Xylene
5. Male, adult, healthy albino rabbits (750-1000g)
6. *o*-Toluidine reagent
7. Wooden rabbit holder
8. Stainless steel feeding needles
9. Distd. water
10. Plastic syringe
11. Syringe
12. Ethyl alcohol
13. Cotton
14. Test sample (crude extract, pure natural product or synthetic compound)

The antidiabetic activity assay involves the following steps:

Preparation of Diabetic Rabbits

1. A group of rabbits is made diabetic by injecting intravenously 150 mg/kg body weight of alloxan monohydrate (**Caution!**)†.

---

†**Caution**! Extreme caution is required to avoid accidental injection in human body

* For further details of all the above mentioned assays see Kiso *et al.*, 1983; Seglen, 1976; Hanks *et al.*, 1949 and Karmen *et al.*, 1955.

2. Eight days after injection, the blood glucose levels of all the surviving rabbits are determined by the *o*-toluidine method (given below).
3. Rabbits with blood glucose levels of 200-500 mg/100 ml are considered as diabetic and employed for the bioassay.

Grouping of Rabbits

1. Normal and alloxan-diabetic rabbits are randomly divided into 5 groups of six animals each.
2. Group 1 serves as a control and receives orally 10 ml of 1% CMC in water.
3. A 0.2 ml sample of blood is immediately collected from group I animals for the blood-glucose determination.
4. Blood samples are also drawn at 5, 10 and 24 hour intervals after the administration of 1% CMC.
5. The animals of groups II, III, IV and V are treated orally with 0.25, 0.5, 1.00 and 1.5 g/kg body weight of test sample suspended in 1% CMC in water, respectively.

Preparation and Adminstrtation of Drug Suspension

1. The amount of test sample required for each rabbit is calculated on body weight basis.
2. The required quantity of extract is suspended in 6 ml of 1% CMC (in water) solution and the final volume made upto 10 ml.
3. The test sample is then administrated orally to each animal by using a stainless feeding-needle on a plastic syringe containing 10 ml of the suspension.
4. The feeding needle is inserted into the stomach through the oesophagus and the plunger pressed slowly and steadily (immediate sneezing and coughing indicates penetration of the needle into the lung; in this case the animal should be rejected and another animal should be taken instead).

Collection of Blood

1. After test sample administration, the animal is held in a wooden rabbit holder and immediately 0.2 ml of blood is collected from an ear vein.

2. Similar samples of 0.2 ml of blood are also collected at 5, 10 and 24 hour time intervals.+
3. After collecting the blood, the pricked side of the ear is rubbed with cotton wool soaked with ethyl alcohol to protect the rabbit against infection.

Determination of Blood Glucose

1. Blood glucose is determined by the method of Fings *etal.* (Fings *et al.*, 1970) using the *o*-toluidine reagent. The *o*-toluidine method is one of the most widely used manual methods.

Statistical Analysis

1. The blood glucose levels in the various groups are expressed in mg/100 ml (Means ± SEM) and the data is statistically analysed by using the variance technique with factorial arrangement.
2. The decrease in blood glucose levels of normal and diabetic rabbits produced by different doses of test sample, found at different time intervals, are compared by using Duncan's Multiple New Range Test (Snedecor, 1965).
3. The standard curve for glucose estimation can be drawn by plotting blood glucose levels of normal rabbits (mg/100 ml) at various time intervals (hours) after oral administration of 1% CMC solution and test sample (0.25, 0.5, 1.0 and 1.5 g/kg body weight) orally, suspended in 1% CMC.

1.9.2 Antidiabetic Activity Assay (GOD-PAP Method)

Diabetes may be defined as a heterogenous and complex metabolic disease (or syndrome) characterized by an inability to utilize carbohydrates. In this state there is an ineffective insulin action because of decreased effectiveness of insulin at the tissue site or a decrease in insulin availabilily. This results in elevated blood sugar levels and concomitant glycosuria. A number of plant extracts and other natural products have been reported to have hypoglycemic activity (Ali *et al.*, 1993).

+ To prevent coagulation of blood, it is sometimes necessary to dampen the rabbit's ear with xylene to promote flow of blood. Xylene causes an inflammatory response, resulting in the blood vessel enlargement and dilation.

Materials

1. Male Wistar strain rats (3-4 months)
2. Streptozotecin
3. D-Glucose
4. Diethyl ether
5. Citrate buffer (0.1 M citric acid, pH 4.5)
6. Peridochrom glucose (GOD-PAP reagent)
7. Distilled water
8. Serum or plasma or blood
9. Centrifuge tubes
10. Centrifuge
11. Spectrophotometer
12. Scissors
13. Water bath
14. Feeding needles
15. Anaesthesia jar
16. Disposable syringes
17. Analytical balance
18. Test tubes
19. Test sample (crude extract, pure natural product or synthetic compound)

1.9.3 Hypoglycemic Activity Assay

Materials

1. Female Wistar rats (200 $\pm$ 25 g)
2. Pentobarbital (50 mg/kg)
3. Standard hypoglycemic medicine (such as glibenclamide)
4. Distd. water
5. Dissection box
6. Esophageal catheter
7. Test sample (crude extract, pure natural product or synthetic compound)

The following procedure is involved in the hypoglycemic activity assay:

1. Female rats are fasted for 20 hours and only given water *ad libitum*.
2. Test sample (5 gm/kg) is administrated orally by means of an esophageal catheter to each rat in a group of rats.
3. Glibenclamide (5 mg/kg) is also administered orally to each rat in another group of rats for comparing the hypoglycemic activity of the test sample.
4. The control rats are dosed with only distd. water.

5. Rats are anesthetized intraperitoneally with pentobarbital (50 mg/kg).
6. The blood samples for glucose determination are collected from the jugular vein at intervals of 30 minutes during 2 hours and are determined with an enzymatic glucose-oxidase method (described below).
7. The data is presented as mean ± SEM.
8. Statistical significance is evaluated by student's *t*-test.
9. The percentage decrease in glycemia in test sample as a function of time is calculated by applying the following formula:

$$\% \text{ decrease glycemia} = \frac{G_o - G_x}{G_o} \times 100$$

where,

G_o = initial glycemia
G_x = glycemia at 30, 60, 90 and 120 minutes (Gx_1, Gx_2, Gx_3, Gx_4) respectively

10. Standard graph showing hypoglycemic activity of the test sample as compared to glibenclamide is drawn by plotting blood-glucose level (mg%) against time (10, 30, 60, 90, 120 min.) (Villar *et al*., 1986).

The steps involved in this bioassay are:

1. Diabetes is induced in adult rats by intraperitonial injection of streptozotecin (65 mg/kg body weight). Diabetic disorders appear after 6 days of injection.
2. After the induction of diabetes the rats are divided into two groups. One group serves as the control and the other as the test group. Animals are then kept for overnight fasting (12 hrs).
3. The test group of overnight fasting rats is fed at 0 minutes with the test sample (250 mg in 2 ml of distilled water) under mild ether anesthesia. The control rats recieve 2 ml of distd. water only.
4. Blood samples are drawn at 0, 60 and 120 minutes under mild ether anesthesia. Before drawing of blood the tails are dipped in warm water (40 ^{o}C) for 30 sec. and the blood is drawn by cutting the tail tip.
5. The blood samples are centrifuged to separate the serum.
6. The blood sugar in the serum is estimated on the same day by the GOD-PAP method given below.

ESTIMATION OF BLOOD SUGAR BY GOD-PAP METHOD

1. 10μl of serum (from each rat) is taken and 1 ml of GOD-PAP reagent is added.
2. 10 μl of deionized water is taken in a test tube (blank experiment) instead of the serum. Both test tubes are treated with GOD-PAP simultaneously.
3. Different concentrations of glucose are prepared in order to plot a standard curve, (e.g. 1 mmolar, 2 mmolar, 4 mmolar, etc.).
4. From each concentration 10 μl is taken in different test-tubes and treated with GOD-PAP reagent simultaneously for the test and blank.
5. The samples, blank and standard, are run in duplicate.
6. All test tubes are incubated at 37 °C for 15 minutes.
7. After 15 minutes, the absorbance is measured by a spectrophotometer at 510 nm.
8. Blood glucose values are calculated from the standard curve of glucose by using the following formula:

$$x = \frac{(y-a)}{b},$$

x = Blood glucose level
y = O.D. of sample
Linear regression constant
b = Linear regression coefficient

9. Significance of the result is calculated student *t*-test.

1.10 DIURETIC ACTIVITY ASSAY

Plant extracts and pure natural products often have a diuretic action which can be screened by a simple *in vivo* bioassay (Kawashima *et al.*, 1985; Schales *et al.*, 1941).

Materials
1. Male Wistar rats (196 ± 1 g)
2. Bicarbonate saline
3. Flame photometer
4. Osmometer
5. Measuring cylinder

6. Metabolism cage
7. Gavage
8. Test sample (crude extract, pure natural product or synthetic compound)

The diuretic activity assay involves the following steps:
1. The rats are randomly divided into 5 groups of 12 each.
2. The animals are fasted overnight and allowed free access to drinking water.
3. Different concentrations of test sample dissolved in a vehicle (sodium bicarbonate) are administered at a volume of 50 ml/kg by gavage to different groups of rats.
4. A control group of rats is administered with a pure vehicle (bicarbonate saline) at a volume of 50 ml/kg by gavage.
5. Individual rats are placed in a metabolism cage. Urine is collected into a graduated cylinder and its volume is recorded at 30 min. intervals for 4 hr.
6. Urinary concentrations of sodium and potassium are determined by a flame photometer. Chloride concentration in the urine is measured by the method of Schales (Schales and Schales, 1941).
7. Urinary osmolality is determined with an osmometer.
8. Significant and dose-related increases in urinary excretion of water (UV) are compared to the vehicle-treated control.
9. Means ± SEM can be presented as figures. Statistical significance can be calculated according to the WSD method for comparison of the data among the means of the experimental group.

1.11 ANTHELMINTIC ACTIVITY ASSAYS

1.11.1 *In-Vitro* Anthelmintic Assay

Materials
1. Helminthis (worm) *Fasciola hepatica, Dicrocoelium dentriticum* or *D. lanceolatum* from infected cattle livers from the slaughter houses
2. Maintenance Medium

Medium 199, Bio-Merieux in conc. sol.	50 ml
Sodium bicarbonate, 5.5% sol.	20 ml
Filtered horse serum	100 ml
Glucose solution 30%	0.5 ml
Distd. water, sufficient to make	500 ml

The pH is adjusted to between 8.2 and 8.5 using 0.1 *N* NaOH
3. Penicillin

4. Streptomycin
5. Sterile sheep erythrocytes
6. Helenin (standard anthelmintic drug)
7. Santonin (standard anthelmintic drug)
8. Dissection microscope
9. Dissection apparatus
10. Syringes
11. Water bath
12. Test sample (crude extract, pure natural product or synthetic compound)

The anthelmintic assay involves the following sequence of steps:

1. The infected cattle livers are obtained from slaughter houses and transported immediately to the laboratory where the worms are dissected out before the temperature of the livers fell appreciably.
2. Care is taken not to injure the fragile worms. In case of multiple infections, *Fasciola* are separated from *Dicrocoelium* (identification characteristics can be obtained from zoology or veterinary medicine departments).
3. The trematodes are placed in a maintenance medium and held in a water bath at 37 °C.
4. At the time of utilization the following chemicals (sterile medium) should be added:

Penicillin	100,000 U
Streptomycin	25 mg
Sterile sheep erythrocytes	1 ml

5. The worms are washed in four changes of sterile medium by slow agitation and decantation. Two or three flukes are then placed in a sterile petri dish containing 50 ml of the medium to which 400,000 units of penicillin are added, and are allowed to remain there for 30 minutes at 37 °C.
6. The experimental flukes are held under these conditions under surveillance for 3 days and any dead specimens are discarded.
7. Fluke vitality is determined by observing their movements with a dissecting microscope.
8. On the third day the surviving flukes are subjected to the various concentrations of test sample and the two reference drugs, helenin and santonin.
9. Various concentrations of the test sample are evaluated.
10. Twenty-four hours after the test sample are added to the petri dishes containing worms and medium, the effect of these products is noted under a binocular microscope (Julien *et al.*, `1985).

1.11.2 *In vivo* Anthelmintic Assay

Materials

1. Seven sheep (~30 kg each) naturally infected with *Dicrocoelium* spp. and having a worm burden which can produce an average of 1200-2700 eggs per gram of faeces.
2. Microlax (Labaz)
3. Compound microscope
4. Test sample (plant extract fraction, pure natural product or synthetic compound)

The protocol which is used for the *in vivo* anthelmintic assay is as follows:

1. Seven sheep are used in this *in vivo* study which lasts for a period of two months.
2. Each sheep, with ears marked, is given one dose of Microlax to facilitate defecation.
3. Faeces are collected and examined on three occasions at five to six day intervals so that the hosts are heavily infected with *Dicrocoelium*.
4. Six sheep are treated with the test sample, administrated by oral route as suspension or aqueous solutions, in different concentrations.
5. Each sheep is given 3 doses of the test sample, at ten day intervals at a rate of maximum 500 mg/kg for the first dose and a maximum of 800 mg/kg for the two subsequent doses. Two infected control sheep do not receive any treatment and act as controls.
6. Coprologic examinations are carried out on all host sheep, 5, 8, 18, 26, 33, 41 and 45 days after the commencement of treatment and the number of eggs per gram faeces is determined.
7. Two infected control sheep which are maintained throughout this two month period should show a nearly constant egg count.
8. The drop in eggs count (eggs per gram of feces) is plotted against days of treatment for different doses (Julien *et al.*, 1985).

1.12 ANTIFERTILITY / ANTI-IMPLANTATION ASSAYS

1.12.1 Antifertility Activity

Materials

1. Female Sprague Dawley rats, 8 weeks old
2. Ethyl alcohol 10%

3. Benzyl benzoate 10%
4. Olive oil 80%
5. Dissection apparatus
6. Animal feed (e.g. Purina rat chow, etc.)
7. Vehicle formulation
8. Test sample (crude extract, pure natural product or synthetic compound)

The antifertility assay involves the following steps:

1. Female rats are selected for the test on the first morning with evidence of positive mating *i.e.* presence of postcoital plugs.
2. The animals showing the presence of postcoital plugs on day-1 are designated as day-1 of pregnancy (hereafter PD_1).
3. The rats are then housed individually in an environmentally controlled room with 12 hrs. of light and 12 hrs. of darkness.
4. They are fed with rat feed (e.g. Purina rat chow) and water *ad libitum.*
5. The pregnant rats are assigned randomly to the control and experimental groups (10 rats in each group).
6. Test samples (in a vehicle) are orally administrated to the rats in the experimental groups numbered from PD_1 to PD_{10}.
7. The rats in the control group are given only the vehicle.
8. The animals are autopsied on PD_{16} (sixteenth day of pregnancy), the number of pregnant animals recorded and the antifertility activity is determined (Kong *et al.*, 1989).

1.12.2 Anti-Implantation Activity

Materials

1. Virgin female rats (Sprague-Dawley) (200 ± 10 g)
2. Tween-80
3. Benzyl benzoate
4. Olive oil
5. Alcoholic Bouni's solution
6. Male rats for mating
7. Ethyl alcohol
8. Test sample (crude extract, pure natural product or synthetic compound)

The steps which are involved in the anti-implantation activity assay are given below:

1. Eight week old virgin female rats are transferred to the holding area and are allowed to settle for one week (fed with Purina

rodent chow and water *ad libitum* in an air-conditioned animal house with light cycle from 06:00 and 20:00).

2. During this time, their estrus cycle is followed.
3. After the second estrus, female rats at proestrus are introduced to the proven male rats.
4. The next morning, the massive presence of sperms can indicate successful mating. These mated female rats are considered to be on pregnancy day 1 (PD_1). Mated rats are numbered and assigned to experimental groups sequentially.
5. The test sample is dissolved in small volumes of ethyl alcohol (final concentration less than 10% in volume) and then diluted with 20% Tween-80 in water. Pure compounds can be dissolved in benzyl benzoate, and then diluted with 4 volumes of olive oil.
6. The formulated test sample is administered to rats intragastrically according to their body weight at around 10:00. The control groups of female rats received the corresponding vehicle (Tween-80 or olive oil).
7. Rats are dosed with test sample from PD_1 (day 1 of pregnancy) to PD_{10} (day 10 of pregnancy) and autopsied on PD_{16} (day 16 of pregnancy).
8. During autopsy, the number of corpora lutea, implantation sites, live or dead/degenerated fetuses and the state of the uterus are noted and recorded (Kong *et al.*, 1985).

1.13 *IN VITRO* ASSAY FOR PLATELET AGGREGATION

The interactions between blood platelets, vascular wall components and coagulation proteins play an important role in major physiological and pathological processes such as haemostasis, thrombosis, atherosclerosis, cancer metastasis and immune disorders. The control of blood platelet aggregation by using natural productsis therefore of potential therapeutic important.

Materials

1. Aggregometer
2. Micropipettes
3. Microsyringes
4. Solutions of aggregating agent such as 5 mM adenosine 5'-diphosphate (ADP) or 2.5 mg ml^{-1} collagen or 0.1 U ml^{-1} thrombin
5. DMSO

6. Fresh human blood (**Caution**!)*
7. Anticoagulant agent such as sodium citrate
8. Disposable large-gauge needle syringes (18/10)
9. Disposable tubes
10. Centrifuge
11. Microscope or automatic cell counter
12. Test sample (crude extract, pure natural products or synthetic compound)

The *in vitro* platelet aggregation assay follows these steps:

Preparation of Platelet-rich Plasma (PRP)

1. Blood is taken from the forearm vein of a healthy donor by a disposable large-gauge needle syringe (**Caution!**) immediately following a moderate venous occlusion in order to limit stasis. The first few milliliters of the blood are rejected.
2. Blood is then transferred (free flow along the side of the tube) into a plastic tube containing 3.8% sodium citrate solution (one ml of sodium citrate solution is required for 9 ml of blood).
3. When the plastic tube is filled with the necessary quantity of blood it is immediately stoppered and inversed gently several times to ensure complete mixing of the blood with the anticoagulant.
4. The blood is then centrifuged 175xg at room temperature for 15 min.
5. The upper layer of platelet-rich plasma (PRP) is carefully removed and kept in a closed plastic tube at room temperature.
6. Platelets are counted under the microscope (or by using an automatic cell counter) and the platelet count is adjusted to 300,000 platelets ml^{-1} with platelet-poor plasma (PPP) which is obtained by centrifugation of the PRP at 10,000xg for 2 min in a microcentrifuge. To ensure a constant platelet count the PRP should be used only during the first 3 h of blood collection.

***Caution** : It should be made sure that the blood donor has not used any drug which is likely to interfere with the platelet functions (specially aspirin or aspirin-containing formulations). Utmost care should be taken during the blood take-out. Vessel damage can lead to the liberation of tissue factors and the generation of thrombin which could activate platelets. Slow blood flow should be performed during the injection, and the movement of the needle inside the vein should be avoided during the blood collection .

Bioassay Procedure

1. The test sample is dissolved in DMSO (1000, 500, 100, 10 ppm). If the test sample is a pure compound a stock solution of 0.1M in DMSO is prepared.
2. One microliter of this test DMSO solution, measured with a precise microsyringe or micropipette, is added to 450 ml of the platelet preparation in an aggregometer under stirring and left to incubate for 30 s, before 50 ml of the aggregating agent is added.
3. The extent of aggregation is compared with that observed in the presence of 1 ml of test DMSO (control).
4. Successive 1/10 dilutions of the DMSO stock solution are assayed until no more inhibition is recorded. Other successive 1/2 dilutions of the lowest active 1/10 dilution are then assayed to determine the minimum inhibitory concentration.
5. The percentage inhibition can be calculated by the following formula:

   % inhibition = ((1- (D/S)) x 100

   where,

   D= the extent of aggregation in the presence of test sample

   S= the extent of aggregation in the presence of solvent alone *i.e.* in control
6. 50% inhibition of aggregation (IC_{50}) can also be calculated from the dose-effect curve (percentage inhibition plotted as a function of the logarithm of the concentration of the inhibitor) by linear interpolation of the linear part of the curve, ranging usually between 25% and 75% inhibition (Beretz *et al.*, 1991).

1.14 ANTI-INFLAMMATORY ASSAY

1.14.1 Rat Paw Edema Assay

Materials

1. Male Wistar rats (Nossan,120-140 g)
2. Indomethacin (Sigma)
3. Carboxymethylcellulose (CMC) (Sigma)
4. Diethyl ether
5. Syringes (0.1 ml, 0.5 ml)
6. Carrageenan (Sigma)
7. Plethysmometer
8. Test sample (crude extract, pure natural product or synthetic compound)

The steps involved in the anti-inflammatory assay are the following:

1. Male Wistar rats are fasted for 12 hr before the experiment.
2. Groups of atleast 5 rats are given 0.5 ml of test sample suspended in 0.5% carboxymethylcellulose.
3. One group of 5 rats is given the standard drug indomethacin (5 mg/kg) in 0.5% CMC.
4. The control group of 5 rats is given only the vehicle (0.5 ml of 0.5% CMC).
5. After one hour of drug administration, rats are lightly anaesthetized with diethylether and paw edema is induced by single subplanar injection of 0.1 ml of 1% carrageenan.
6. Paw volumes are measured using a water plethysmometer immediately before the injection of carrageenan and at hourly intervals for 5 h thereafter.
7. The volume of edema is expressed for each rat as the difference before and after the injection of carrageenan.
8. The percent inhibition of edema is calculated for each group (test sample-treated group and standard drug-treated group) versus its vehicle-treated control group.
9. Data are analyzed using unpaired student's *t*-test and a $p < 0.05$ (probability) is taken as significant (Aquino *et al.*, 1991).

1.15 IMMUNOMODULATING ASSAY

1.15.1 Hemolytic Plaque Assay

Materials

1. Balb/c female mice (six week old)
2. Saline (0.85% sterile NaCl)
3. Sheep red blood cells (SRBC)
4. Syringes
5. Dissection apparatus
6. Wire-gauze mesh
7. Petri dishes
8. Eagle's medium
9. Incubator
10. Dissecting microscope
11. Glass slides
12. Agar
13. Trypan blue
14. Guinea pig complement
15. DMSO
16. Humidity chamber

17. Test sample (plant extract, pure natural product or synthetic compound)

The hemolytic plaque assay involves the following steps:

1. Different groups of six week old female mice are first injected intraperitonially (IP) with 10 mg of the test sample.
2. On the same day mice are immunized with 0.1 ml of SRBC diluted to 10% concentration with 0.85% sterile NaCl as antigen.
3. After 24 hr. these mice are given a second IP injection of 10 mg of the test sample.
4. Two groups of control animals are included. One group is given only 0.1 ml IP injection of SRBC as antigen while the other receives 0.1 ml DMSO.
5. On days 4 to 5 of the SRBC immunization, the mice are killed by cervical dislocation and the spleens are removed.
6. The spleens are then processed into single cell suspension with the help of a wire-gauze mesh in a Petri dish.
7. The cell suspension in Eagle's medium is centrifuged at 200 RPM for 5 min.
8. The pellet is washed twice and finally suspended in 1 ml Eagle's medium.
9. Two tenfold dilutions are made and the cells are counted in a hemocytometer. The viability of the cells is determined by trypan blue exclusion.
10. The cells are kept in ice until used.
11. A mixture containing 0.3 ml plaquing medium (2% agarose in Eagle's medium, 0.15 ml washed 10% SRBC and 0.1 ml of one of the spleen cell dilution) is spread onto 1% agarose coated slides and allowed to solidify at ambient temperature for a few minutes.
12. The slides are then incubated in a humidity chamber for 1 hr at 37 ^{o}C.
13. After incubation, the slides are placed face down and reincubated with fresh Eagle's medium and guinea pig complement (1:20).
14. After 30-45 minutes of incubation, the slides are observed under a dissecting microscope and the splenic plaque forming cells (SPFC) are counted (Kazmi *et al.*, 1990).

1.16 ANTIEPILEPTIC (ANTICONVULSANT) ASSAY

Epilepsy is a common disorder in which nerve cells of the brain from time to time release abnormal electrical impulses. These cause temporary malfunction of the other nerve cells of the brain, resulting in alteration of, or complete loss of consciousness.

A carefully collected data from international survey indicates that about 1 adult in 200 suffers from recurrent epilepsy. The synthetic antiepileptic medicines available do not cure the disease completely and often have severe side effects. There is therefore a need to discover new antiepileptic agents from traditional herbal therapies.

1.16.1 Pilocarpine Model

Structural damage of the rat's brain, induced by systemic administration of the cholinergic agents such as pilocarpine, may lead to chronic spontaneous epileptic seizures in survivors. Pilocarpine is a potent muscarinic cholinergic agonist, which is capable of producing a sequence of behavioral alterations, including staring spells, facial automatisms and motor limbic seizures, that develop over 1-2 hr and build progressively into limbic status epilepticus. After a silent period (mean 11-44 days) animals enter into chronic period characterized by spontaneous recurrent seizures (SRS) which continue for the life of the animals. This condition provides an excellent experimental model for the search of antiepileptic (anticonvulsant) natural products.

Materials

1. Thirty-two adult Wistar rats (250-280 gm).
2. RM polygraph
3. Bipolar twisted wire electrodes (tip diameter 100 μm, interelectrode distance 500 μm)
4. Sodium pentobarbitol
5. Pilocarpine hydrochloride (convulsant agent)
6. Methyl scopolamine nitrate
7. Video supported EEG monitoring system (optional)
8. Light microscope
9. Acetic acid 10%
10. Formaldehyde 10%
11. Methanol 80%
12. Paraffin
13. Glass slides
14. Cresyl violet
15. Saline
16. Tween 80,3%
17. Plexiglas observation compartments (40x25x17cm)
18. Plexiglas recording compartments (30x30x45 cm)
19. Syringes

20. Jeweller screws for surface EEG
21. Standard antipileptic drugs (phenobarbitone, phenytoin, carbamazepine, valproic acid and ethosuximide)
22. Test sample (crude extract, pure natural product or synthetic compound)

The method used in the assay is as follows:

1. Rats are kept under environmentally controlled conditions (6:00 AM/6:00 PM light/dark cycle: 22-25 ^{o}C) with free access to food and water for 7 days before the experiment.
2. Pilocarpine hydrochloride is freshly dissolved in saline and administered in doses of 380 mg/kg.
3. Methylscopolamine nitrate is injected subcutaneously (1 mg/kg) 30 min prior to pilocarpine administeration in order to minimize the peripheral cholinergic effects.
4. Pentobarbital (30 mg/kg *i.p.*) is injected 6 hr after the start of status epilepticus to decrease the otherwise high mortality rate.
5. To establish baseline seizure frequency during SRS period, animals are injected with solvents (saline or Tween-80, 3%) for 2 weeks and the numbers of SRS per 5 days of each week are noted.
6. The animals are then divided into three groups and observed and treated for 5 days per week for two weeks under observation.
7. The control group continues to recieve solvents (saline, or tween-80, 3% *i.p.*) only. The second group recieves one or more of the following: phenytoin 100 mg/kg, carbamazepine 120 mg/kg, valproic acid 450 mg/kg and ethosuximide 400 mg/kg *i.p.* The third group receives the test sample in various doses *i.p.*

Behavior

1. Behavioral assessments are carried out in observation compartments (40x25x17 cm) built of plexiglas. Each animal is habituated to these compartments for 30-45 min prior to administration of pilocarpine.
2. After habituation, rats are removed and injected i.p. with pilocarpine and rapidly returned to the experimental cages.
3. Following the status period (60-90 days), the surviving rats are continuously monitored for the following 120 days.
4. Animals presenting 3 or more seizures per five days of observation for 2 weeks during SRS period are included for pharmacological studies.

Surgery and EEG Recording

1. For EEG recording (two per group of animals) bipolar twisted wire electrodes are positioned stereotaxically under sodium pentobarbital anesthesia (50 mg/kg *i.p.*) in the right dorsal hippocampus (A= 3.5, L= 2.5, D= 3.5).
2. Surface recordings are made from jewellers screws.
3. Signals under investigation are amplified by a polygraph (time constant 0.0-3 sec., high cut off filter 30 Hz).
4. EEG recordings are performed 5-7 days after surgery.
5. Recordings are carried out in a plexiglas compartment.
6. Before EEG recording, animals are placed individually in the recording compartment and allowed 20-30 min for habituation to the experimental setup.
7. After the period of habituation, baseline EEG recordings are made continuously during the status period ranging from 6-8 hr following pilocarpine injection.
8. During the chronic period animals are submitted to continuous EEG recording (24 hours/day) in order to observe the influence of light-dark cycle over the seizure frequency.
9. Videotaping (optional, 24 h/day for 120 day) is carried out of spontaneous seizures in animals. The monitoring system provides complete information about the behavior of animals during the chronic period.

Morphological Examination

1. 120-125 days after the administration of pilocarpine, the rats are anesthetised with an over-dose of sodium pentobarbital and perfused with a fixative containing 10% acetic acid, 10% formaldehyde and 80% methanol.
2. The rat brains are allowed to fix *in situ* at 4 ^{o}C for 24 h and then removed and processed for paraffin embedding.
3. Serial coronal sections of the whole brain are cut (10-mm thick) and every tenth section is mounted on a glass slide, stained with cresyl violet and examined by light microscope.
4. The correct location of implanted deep electrodes is histologically determined in cresyl violet-stained serial sections by light microscopy.

Results Analysis

1. Behavioral alterations in control and test groups of animals are observed throughout the status and chronic periods.

2. Antiepileptic activity of test sample is measured as % reduction in seizures in test group as compared to control group of animals.*

1.17 ANALGESIC ASSAYS

1.17.1 Writhing Syndrome Method

Materials

1. Acetic acid
2. Mice (20-30 g)
3. Syringes
4. Stop watch
5. Test sample (crude extract, pure natural product or synthetic compound)

The following steps are involved in the writhing syndrome assay:

1. Writhing syndrome is introduced by injecting (intraperitoneally) 3% acetic acid to the mouse at a dose level of 300 mg/kg (10 ml/kg).
2. After 30 and 60 min., 1, 5, 10, 50 and 100 mg of the aqueous solution of the test sample is injected intraperitoneally into the animal.
3. The number of writhes are noted for a 10 min period, between 5 and 15 min after the acetic acid injections (Chapman *et al.*, 1982).

1.17.2 Hot Plate Method

Materials

1. Hot plate
2. Mice (20-30 g)
3. Syringe
4. Standard analgesic morphine
5. Test sample (crude extract, pure natural product or synthetic compound)

The hot plate method consists of the following steps:

1. Mice are injected with 1, 5, 10, 50 and 100 mg aqueous solutions *i.p.* of the test sample.

---

* For further detail see Leitz *et al.*, 1990;

2. These animals are then left on hot plates (53.5 ^oC) 30 and 60 min after the injections.
3. The time when the animals lick either of their hindpaw is accepted as the end-point.
4. The cut-off time is 60 sec.
5. Morphine can also be injected to the animals as a standard analgesic to compare the activity (Benedek *et al.*, 1948).

1.18 GASTROPROTECTIVE/ANTIULCER ASSAYS

1.18.1 Gastroprotective Assay

Materials
1. Female cross-bred Albino rats
2. Absolute ethanol
3. 10% Formalin
4. Ether
5. Dissection apparatus
6. Dissection microscope
7. Test sample (plant extract, pure natural product or synthetic compound)

The following sequence of steps is involved in the gastroprotective assay:
1. The rats are randomly divided into five groups:
   GP I: Treated with distilled H_2O (control).
   GP II: Treated with extract containing 500 mg test sample/kg.
   GP III: Treated with extract containing 375 mg test sample/kg.
   GP IV: Treated with extract containing 250 mg test sample/kg.
   GP V: Treated with extract containing 125 mg test sample/kg.
2. Food is withheld from the rats for 36 hrs. and water for 24 hrs. before starting the experiment.
3. 1 ml of the test sample is orally administered to each animal.
4. 30 min. later 1 ml of absolute ethanol is given orally to induce gastric lesions in each animal
5. The animals are killed after one hour by an overdose of ether.
6. Their stomachs are taken out, gastric contents are aspirated, and the stomachs inflated with 10% formalin (v/v) and then immersed in 10% formalin for 6 min.
7. The stomachs are then examined for linear hemorrhagic lesions in the glandular region.

8. The length (mm) of each lesion is determined at x10 magnification with a pair of dividers and each length is summed per stomach.
9. The results are given as mean ± S.E.M. (Yamahara *et al.*, 1988).

1.18.2 Anti-ulcerogenic Assay-1

Materials

1. Male Wistar rats (160 - 180 g)
2. Physiological saline (0.86% NaCl)
3. Ulcerogenic compounds (phenylbutazone, serotonin creatinine sulfate, EtOH, aspirin or indomethacin)
4. Wire cage (4.5 x 4.5 x 18 cm)
5. Formalin 1%
6. Distd. water
7. Dial caliper
8. Test sample (crude extract, pure natural product or synthetic compound)

The antiulcerogenic assay-1 consists of the following steps:

1. Rats are fasted on water for 48 hr. prior to each treatment in which a group of 7-10 rats are used.
2. Test sample, dissolved in physiological saline, is administered to rats by different routes (i.v., p.o., i.d.) 30 min. prior to the ulcer-inducing treatment of animals.
3. The control group of 7-8 rats are administered ulcerogenic agent and physiological saline only.
4. For inducing ulcerogenesis in rats phenylbutazone (250 mg/kg) or serotonin creatinine sulfate (30 mg/kg) are administered to rats orally and subcutaneously, respectively (ulcerogenesis can also be induced by oral administration of 50% EtOH, 1 ml/rat, or aspirin 100 mg/kg).
5. The animals are sacrificed by the inhalation of ether, 5 hr. after each treatment, and their stomachs are taken out for pathological observations.
6. Each ulcerated stomach is fixed with 1% formalin for 3 min. then incised along the great curvature, and rinsed with physiological saline to remove the gastric contents and blood clots.
7. The areas and the lengths of hemorrhagic lesions induced by serotonin and other ulcerogens, respectively, are measured using a dial caliper, and the sum of measurements for each animal is referred to as the ulcer index.

8. The antiulcerogenic activity of the test sample is evaluated by comparison of the mean values of the ulcer index between the test sample treated group and the control (physiological saline) groups.
9. Data obtained from the experiment is expressed as mean ± S.E. Statistically significant differences between the treated and the control groups are tested by the student's *t* -test.
10. The potency of the test sample is expressed in terms of the 50% inhibition dose (ID_{50}) (Tanaka *et al.*, 1989).

1.18.3 Anti-ulcerogenic Assay-2

Materials

1. Male Wistar rats (200 g)
2. Hydrochloric acid
3. Ethyl alcohol
4. Formalin solution (2%)
5. Dissection apparatus
6. Dissection microscope
7. Test sample (plant extract, pure natural product or synthetic compound)

Another procedure for the anti-ulcerogenic assay is the following:

1. The rats are divided into groups of 5-7 rats.
2. After 24 hr. of fasting, they are given the test sample dissolved in methanol or acetone, orally (300-1000 mg/kg).
3. One hr. after the oral administration, 1 ml of HCl/ethanol solution (150 mM HCl in 60% ethanol) is administered orally to each rat.
4. One hr. after the administration of the HCl/EtOH solution the rats are killed by ether and the stomach excised.
6. Following treatment of the stomach in 2% formalin solution, the lesion index is calculated as the sum of the lengths (cm) of lesions found in the mucosal membrane of the fundus region (Yamahara *et al.*, 1988).

1.19 RADIOLABELLING BIOASSAYS

1.19.1 Mitogenic Activity Assay

Materials

1. Liquid scintillation counter
2. Multiple cell harvester

3. [^3H] - Thymidine (specific activity 185 GBq; 5.0 Ci/mmol)
4. Centrifuge (300 rpm)
5. Sterile nylon wool column
6. Phosphate buffered saline (PBS)
7. RPMI 1640 medium
8. Toyo GC 50 glass fiber filters
9. Trypan blue dye
10. 96-well microtiter plates
11. Humidified incubator
12. Male 5-7 week old BALB/c, BALB.c-nu/nu C3H/HeJ mice
13. Leuco PREP
14. Lympholyte-M
15. Trypsin-chromatrypsin inhibitor
16. Human serum type AB
17. Pronase
18. Bacto phytohemagglutinin P (PHA)
19. Sterile distd. water
20. Penicillin
21. Streptomycin
22. Test sample (plant extract, pure natural product or synthetic compound)

The mitogenic activity assay involves the following steps:

Preparation of Test Sample

1. Water extract or water soluble natural product is filtered and used as standard test sample solution.
2. The standard test sample solution is diluted with RPMI 1640 medium supplemented with 10% human serum (type AB), 100 U/ml of penicillin and 100 μg/ml of streptomycin (complete RPMI medium) to make 1, 1/10, 1/100 and 1/1000 dilutions.

Preparation of Lymphocyte Suspension

1. Human lymphocytes are separated from peripheral blood by gradient differential centrifugation with Leuco prep at 3,000 rpm for 30 min.
2. T and B cells are isolated by the following method:
   - 2.1 The sterile nylon wool columns are rinsed with phosphate buffered saline (PBS) and with complete RPMI medium.

2.2 Cell suspensions are loaded onto the column and washed into the nylon wool with RPMI medium.

2.3 The column is left for 40 min. at 37°C and then slowly washed with RPMI medium.

2.4 The effluent from the column contains the T cells while the nylon well adherent cells are collected for the B cells.

3. After washing once with PBS and twice with complete RPMI medium, the lymphocytes are resuspended in the same medium at the concentration of 1 x 10^6 cells/ml.
4. Mouse lymphocytes are isolated from the spleen by gradient differential centrifugation with lympholyte-M at 1,500 rpm for 30 min. They are washed in the same way as human lymphocytes and are suspended in the same medium at a concentration of 5 x 10^6 cells/ml.
5. The viability of lymphocytes, determined by the trypan blue dye exclusion test, is more than 99% in all the preparation.

Determination of Mitogenic Activity

1. Lymphocyte suspension (100 ml) and test sample (100 ml) (1 to 1000 dilution) are mixed and incubated in 96-well microtitor plates in humidified incubation with 5% CO_2-95% air at 37 °C for 48 h (human lymphocytes) or 66 h (mouse lymphocytes).
2. [^{3}H]-Thymidine (9.25 KBq; 0.25 mci/well) is then added to the well, and incubated for another 24 h or 6 h for human or mouse lymphocytes, respectively.
3. The incorporation of [^{3}H]-thymidine by lymphocytes is determined by collecting the cells on Toyo GC 50 glass fiber filters using a Labo-Mash multiple cell harvester.
4. Radioactivities on the filters are measured by a liquid scintillation counter in a toluene base scintillation fluid.
5. Stimulation index (S.I.) is measured as follows using the mean values of three to five independent experiments performed in triplicate:

$$\text{S.I.} = \frac{[^3\text{H}]\text{-incorporation (dpm) of cultures treated with test sample solution}}{[^3\text{H}]\text{-incorporation (dpm) of control cultures}}$$

6. The statistical significance of the results is evaluted using student's *t*-test (Tachibana *et al.*, 1992).

1.20 ANTI-EMETIC ASSAY

Materials

1. Emetic agent ($CuSO_4.5H_2O$)
2. Leopard (*Pana nigromaculata*) and Ranid (*Pana japonica*) frogs of either sex
3. Tween 80
4. Earthworm
5. Test sample (crude extract, pure natural product or synthetic compound)

The bioassay method employed is as follows:

1. The frogs are divided into groups consisting of 6 each
2. Each frog is orally fed with earthworms 3 hr before the experiments.
3. The test sample is suspended in 5% Tween 80 and administered orally at doses of 10-1000 mg/kg body weight.
4. The control group of frogs receives only 5% Tween 80 solution
5. The frogs are allowed to stand for 30 min to stablize their conditions.
6. The copper sulfate pentahydrate (emetic agent) is administered orally and the first emesis (emetic latency) is recorded for 80 min.
7. The results are judged by the prolongation of the emetic latency.
8. A significant prolongation of emetic latency is the sign of an anti-emetic action of the test sample.
9. All numerical data are expressed as the means of ± S.E. The statistical significance of the difference is determined by an unpaired student's *t*-test (Kawai *et al.*, 1994).

PART-B

High-Throughput Screening

2.1 INTRODUCTION

Extracts from natural product sources have been a valuable source of molecular diversity in many drug discovery programs and several therapeutically important drugs have been isolated from natural products. However, the screening of such materials can be very complicated due to fact that they contain a complex mixture of secondary metabolites. Success in identifying new therapeutics using *in vitro* receptor and enzyme assays is largely dependent upon the proper design and validation of these screening assays. Another key to success is the design of assays with higher throughput so that a large number of extracts can be screened quickly and cost effectively. The advent of combinatorial chemistry methods has also enhanced the importance for high throughput screening methods for testing for bioactivity. This section contains a detailed description of the design and validation of receptor and enzyme assays used to screen combinatorial chemistry reaction products and natural product extracts. In addition, new assay technologies designed to increase assay throughput will also be presented.

2.1.1 Characteristics of an Ideal Screening Assay for Screening Natural Product Extracts

There are several important characteristics associated with an ideal primary screening assay used for screening natural product extracts and combinatorial chemistry reaction products. Table **2** details these characteristics. Each of these desirable characteristics are discussed in detail below.

Table 2: Characteristics of an Optimal Screening Assay

- Biochemical or cell-based
- Assay format: 96- or 384-well microtiter plate, amenable to automation
- Reagent source: human recombinant reagents
- Assay signal: high and reliable signal-to-noise ratio
- Assay incubation: 30-60 minutes at ambient temperature
- Assay manipulations: minimal
- DMSO compatible
- Source material compatibility: with all types of natural product extracts and other chemical diversity

2.1.1.1 Biochemical or Cell Based Assays

Generally, there are two types of assays used for high throughput screening: biochemical or cell-based assays. Biochemical based assays utilize partially purified or purified human receptors or enzymes as assay reagents. Advantages of using biochemical assays are that continual cell culture is not required, cell permeability barriers are not present (for intracellular targets), and determination of the mechanism of action of an active extract or compound identified is generally uncomplicated. Potential disadvantages of biochemical assays are that the assay conditions do not reflect the physiological environment present in cells and often they do not permit development of functional assays.

Similarly, there are several advantages and disadvantages to using cell-based assays in screening combinatorial chemistry reaction products or natural product extracts. One advantage is that several interesting molecular targets may be contained in one assay. For example, a cell-based assay involving measurement of IL-2 production by stimulated T cells would contain several molecular targets that, if inhibited, would lead to reduced IL-2 production. Some of these molecular targets would include the tyrosine kinases $p56^{lck}$, $p59^{fyn}$, ZAP70, the tyrosine phosphatase CD45, as well as other targets that are involved in T cell signal transduction leading to IL-2 production (Crabtree *et al*, 1994). Another advantage of using cell-based assays is that functional assays can be developed. An example of a functional cell-based receptor assay for G-protein coupled receptors would be one in which levels of the appropriate second messenger such as cAMP or intracellular calcium are measured. The advantage of using a cell based functional assay is that the assay can be used to distinguish agonists and antagonists. Another benefit of using cell-based assays is that reagents do not need to be purified, and cells just simply need to be grown. One potential caveat of cell based assays, which involve complex signal transduction pathways leading to an assay endpoint signal, is that determination of the locus of action of an active reaction mixture or extract may be complicated. One of the paramount concerns of screening natural product extracts in cell-based assays is the potential for a high incidence of false positive active extracts due to nonspecific cellular cytotoxicity.. Because of this, active extracts identified in cell-based assays should be routinely evaluated for nonspecific cytotoxicity.

2.1.1.2 Assay Format

Another important consideration for design of receptor and enzyme screening assays is assay format. The assay should be formated in a 96-

well or 384-well microtiter plate format. Once in this configuration, microtiter-plate based liquid handling devices and endpoint plate readers can be used to dramatically increase assay throughput. In addition, microtiter plate-based robotics systems can also be utilized. Thus, the instrumentation that has been used for high throughput screening of small organic molecules can also be applied to high throughput screening of natural product extracts. Logically, the more extracts screened per assay in a natural product drug discovery program, the greater probability will be the that a valuable secondary metabolite will be identified. New technologies currently being developed for high throughput screening which will result in even greater throughput, such as assay miniturization, should also be applicable to screening natural product extracts. One consideration for high throughput screening of natural products is that active extracts will accumulate, waiting for chemical isolation and identification of the active secondary metabolite. Natural product chemistry is typically the more time consuming component of the natural product discovery process.

2.1.1.3 Reagent Source

Receptor and enzyme screening assays should utilize cheap, consistent, and reliable sources of reagents. Traditionally, animal tissues have been used as a source of reagents for screening assays but this source has become less desirable due to expense and "animal rights" issues. More importantly, the pharmaceutical industry has recognized from past experience that high sequence homology for receptors and enzymes isolated from different species does not necessarily mean that human proteins will share similar pharmacological characteristics. Ideally, a primary screening assay used in a drug discovery program should utilize human reagents. Human cell lines expressing receptors or enzymes of interest may serve as a good reagent source if cells contain a single pure population of the desired reagent and if purification is reasonably simple. However, it is common that structurally and pharmacologically related receptors (receptor subtypes) or enzymes (isoforms) may be present in any given cell which can complicate the screening assay.

The most desirable source of reagents for screening assays are human recombinant proteins. The advantage of producing recombinant proteins is that they serve as a cheap, reliable and pure source of reagents. Another important advantage of using recombinant proteins is that desirable physical characteristics can be introduced. Several expression systems can be utilized including mammalian cells, bacteria, yeast and insect cells. Mammalian cells are the best expression choice if functional assays are developed because they will usually contain the appropriate signal transduction machinery required. One concern with production of recombinant proteins in non-mammalian expression systems is whether they will undergo the same post-translational modifications that occur physiologically. Because of this concern, the characteristics of proteins produced in these surrogate expression systems should be closely evaluated to make sure that they are suitable.

2.1.1.4 Assay Signal-to-noise Ratio

Primary assays used for screening combinatorial chemistry reaction products and natural product extracts require a high signal-to-noise ratio. This is especially relevant for screening crude natural products because of the high inherent noise associated with screening these complex mixtures of secondary metabolites. In developing receptor and enzyme assays, each assay variable should be evaluated to determine if it may influence the levels of assay noise. Variables found to increase signal and reduce noise should be adjusted to maximize the signal-to-noise ratio. For example, assay noise in radioligand receptor binding assays typically relates to nonspecific binding of radioligand. There are several assay variables that can strongly influence nonspecific binding including radioligand and membrane concentrations present in the assay. For enzyme assays, assay noise can be a contribution of many different factors and like receptor assays, it can often be reduced by manipulation of different assay parameters.

2.1.1.5 Assay Incubation

The length of assay incubations is a critical consideration for developing both receptor and enzyme assays. For receptor assays, incubations must be of sufficient duration to achieve steady state binding, which is simply the period in which the rate of ligand association with the receptor equals the rate of ligand dissociation from the receptor. Ideally, once equilibrium is reached, assay plates can set without being further processed if the receptor and the radioligand are stable. This is an advantage when screening large numbers of plates because the binding reaction does not have to be terminated at precise intervals. Incubations for enzyme assays are more restricted and should be performed under initial rate conditions where the reaction is of the first order with respect to substrate concentration. Since the enzyme reaction is linear under initial rate conditions, assays have to be precisely initiated and terminated at discrete intervals.

Incubation temperature is another variable to consider for both receptor and enzyme assays. Reaction rates will be more rapid at higher assay temperatures. However, incubations performed at ambient room temperature are more practical because incubators are not required. Although most reactions can be performed at room temperature, some exceptions exist and will be discussed in the following sections.

2.1.1.6 Assay Manipulations

Simple assays in which minimal assay manipulations are performed will have the greatest screening throughput. Some assays involve too many manipulations to make them practical for screening large numbers of natural product extracts. For receptor binding assays, there is now a shift from traditional filtration binding assays to homogenous assays in which reagents are added, incubations are conducted, and plates are read in endpoint readers. Thus, the laborious steps of filtration to resolve free and bound radioligand can be omitted. Similarly, homogenous enzyme assays are being developed which allow omission of laborious separation steps to resolve substrate from product. The main point is that an ideal assay will have a minimum number of assay manipulations, thereby allowing greater assay throughput.

2.1.1.7 DMSO Compatibility

Usually, natural product extracts are suspended in DMSO for screening in primary assays. An optimal screening assay should tolerate high concentrations of DMSO to maximize the probability that secondary metabolites in an extract will remain soluble during the screening assay incubation. For any assay, it is critical that the DMSO tolerance be evaluated carefully. Most receptor binding assays utilizing membrane preparations will tolerate up to 10% DMSO. Most enzyme assays will also tolerate relatively high levels of DMSO. Cell based assays have a much lower compatibility and usually will not tolerate final assay concentrations of above 0.5% DMSO.

2.1.1.8 Screening Source Material Compatibility

Most natural product drug discovery programs consist of screening of different extracts from a variety of source materials including actinomycetes, fungi, plants, insects, and marine organisms. In addition, for any organism, it is common to prepare different types of chemical extracts for screening. For any screening assay, it is important to screen a relatively large sample of each type of extract during assay validation to evaluate its compatibility with a given assay. Many assays have different capacities for different types of extracts and source materials. Generally, determination of the optimal final extract concentration of any given type of natural product extract is accomplished by testing various dilutions of extracts from several organisms and determining the concentration which gives an acceptable active hit rate, which is generally between 0.1 and 1%. During the lifetime of a screening assay, it is common that the final assay concentration of a given set of extracts may be varied to achieve different hit rates.

2.2 ENZYME ASSAYS

2.2.1 Introduction

Certain enzymes are important molecular targets for drug discovery and many drug discovery programs include enzyme targets in primary screening assays. Generally, these assays utilize partially purified or purified human enzymes and involve the measurement of product formation by a variety of assay methods including radiometric, colorimetric, and fluorometric assays. The following sections detail important considerations for developing and validating enzyme assays for synthetic compounds and screening natural product extracts. In addition, an example of development and validation of a specific enzyme assay is presented to serve as an example.

2.2.2 Considerations for Development of Enzyme Assays

The following represent the important assay variables that should be considered in developing high throughput enzyme assays for screening combinatorial chemistry reaction products:

- Source of enzyme
- Final assay concentration of enzyme
- Selection of substrate and final assay concentration
- Cofactor requirements and final assay concentrations
- Selection of assay endpoint
- Assay pH
- Assay temperature
- Inhibitor profiles
- DMSO compatibility
- Extract compatibility

Before attempting to develop an enzyme assay, it is important to learn as many details concerning the mechanism of the enzyme reaction as possible. Important information to evaluate would include whether an enzyme reaction involves a single or multiple substrates, the mechanism of catalysis, what cofactors are required, and factors which regulate enzymatic activity. All of this information will have impact not only on the proper design of the enzyme assay but also in establishing the mechanism of action of an inhibitor found during screening.

2.2.2.1 Selection of Enzyme Source

One of the first considerations for developing an enzyme assay concerns the selection of a viable source of enzyme. If possible, enzymes used in assays for drug discovery should come from human sources. Most enzymes isolated from different species will be highly homologous but may not necessarily share all the same characteristics of the human enzyme. To insure that the assay is relevant to finding inhibitors of a human enzyme of interest, human enzymes should be utilized. Human tissues can serve as a source for human enzymes but they are generally difficult to obtain. Human cells or cell lines are a more practical source of enzymes. In some instances it may be necessary to use an enzyme from a non-human source if a good human source is not available. If this is the case, it will be important to evaluate any active secondary metabolite identified in the primary assay for activity against the corresponding human enzyme.

Once a source of enzyme is identified, the next task is to purify the enzyme in sufficient quantities for screening. Purification of some enzymes from cells can be very laborious and yields can be low either due to the presence of low amounts of enzyme or enzyme degradation during purification. In some instances, cellular expression of an enzyme can be increased prior to purification. An example is the enhancement of the expression of inducible NO synthase when mouse macrophage-like RAW 264.7 cells are cultured in the presence of mouse γ-interferon and LPS. Another issue concerning purification of an enzyme from cells concerns purity. Many enzymes exist as multiple structurally related isozymes. A cell line expressing the particular isozyme of interest needs to be identified and a means to resolve it from other isozymes, if present, needs to be established. In many cases, the cell culture will need to be scaled up to provide enough cells for isolation of the enzyme of interest.

A more convenient method for producing enzymes for screening assays is to clone and express the human enzyme of interest in an appropriate expression system. Cells expressing recombinant human enzyme can generally be used to purify large amounts of pure protein cheaply. It is also possible to engineer certain groups into recombinant enzymes so that they can be more easily purified. One example is the introduction of a C-terminal histidine residue into a cytosolic enzyme which enables a simple one-step purification using an iron chelating column. Several different expression systems can be considered for expressing human recombinant proteins including mammalian cells, bacteria, insect cells, or yeast. In selecting an expression system, it is important to establish that the expression system does not express a related endogneous protein and that the expressed protein will undergo the same post-translational modifications that occur physiologically. There are a number of different reagent suppliers that sell recombinant enzymes that can be used for screening assays.

2.2.2.2 Selection of Enzyme Concentration

The final concentration of an enzyme included in a screening assay is another important consideration in assay development. Since purified human enzymes may be expensive and laborious to isolate, the amount of an enzyme included in an assay should be minimized but still be sufficient to generate the desired signal-to-noise ratio. This variable is easily examined by performing an experiment in which variable amounts of enzyme are added to the assay. Since many purified enzymes may be labile during storage, appropriate storage conditions should be established. In addition, stability of the enzyme should be evaluated periodically during periods of storage.

2.2.2.3 Selection of Substrate and Substrate Concentration

a) One Substrate Reversible Enzyme Assays

The simplest enzyme reaction is one in which a single substrate is involved and the conversion to product is reversible. Such a simple enzymatic reaction is shown below:

$$A \underset{}{\overset{E}{\rightleftarrows}} P$$

In general, the substrate selected for a single substrate assay should possess a high affinity (low K_m) for interaction with the enzyme. This will allow the use of low amounts of the substrate in the assay as well as provide a sensitive assay to identify inhibitors that compete with the substrate for the enzyme active site. Typically, physiologically relevant substrates are used which will generally have a low K_m for enzyme. However, some assays utilize synthetic peptides or derivatized peptide substrates. It is important that these derivatized substrates possess relatively high affinity for enzyme so that the enzyme assay is sensitive in identifying active secondary metabolites that may be present in low concentrations.

The K_m of an enzyme for a particular substrate can be easily determined experimentally from evaluating the relation of the concentration of substrate ([S]) to the initial reaction rate (v_o). Such a relationship is shown in the Figure **1**.

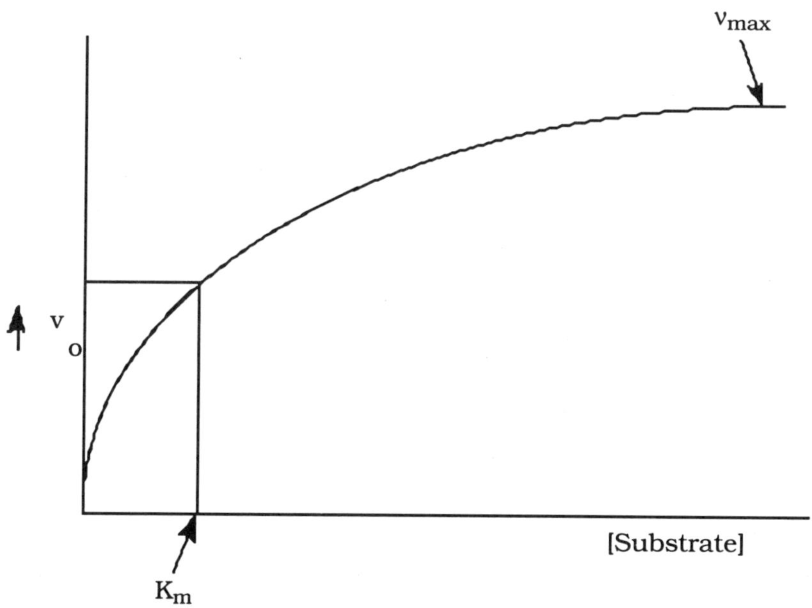

Figure 1: Relation of substrate concentration to rate of a simple one substrate enzyme reaction

At low substrate concentration, the initial reaction velocity is nearly proportional to the substrate concentration and the reaction is thus approximately first order with respect to substrate. However, as the substrate concentration is increased, the reaction rate becomes essentially independent of substrate concentration and approaches a constant rate. In this range of substrate concentration, the reaction becomes zero order with respect to substrate concentration since the enzyme is saturated with substrate. According to the Michaelis-Menton equation for a single substrate reaction, $v_o = V_{max}$ [S]/ K_m + [S], K_m where is equal to the concentration of substrate at which the initial reaction velocity is half the maximal velocity.

The Michaelis-Menton relationship can be algebraically transformed into other graphical forms that can provide important kinetic information. One example is the Lineweaver-Burke plot in which the reciprocal of the reaction velocity ($1/v_o$) is plotted against the reciprocal of the substrate concentration (1/[substrate]). Such a plot in shown Figure **2** for a single substrate enzymatic reaction.

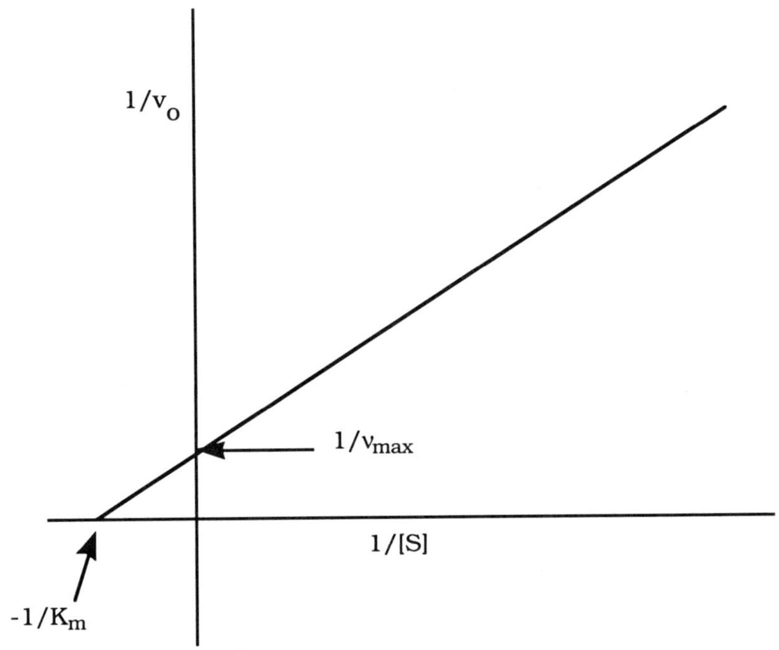

Figure 2: Lineweaver-Burke plot.

The y-intercept is equal to $1/V_{max}$ and the x-intercept is equal to $-1/K_m$. Another useful transformation is the Eadie-Hofstee plot in which v_o is plotted against v_o/[S]. Such a plot is shown in Figure **3**.

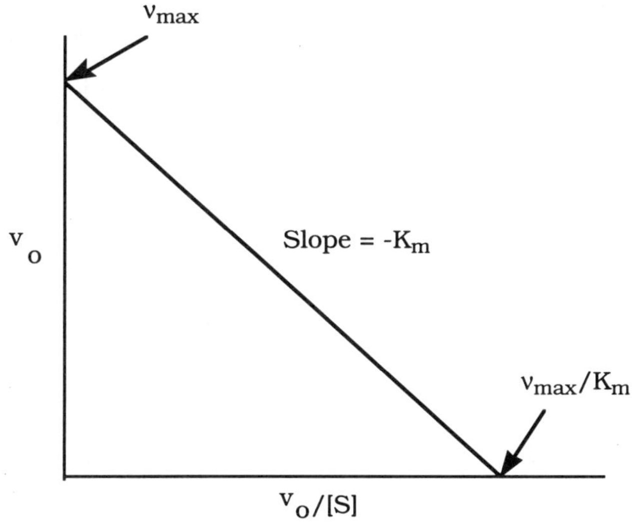

Figure 3: Eadie-Hofstee plot.

The y-intercept is equal to V_{max}, the x-intercept is equal to V_{max}/K_m, and the slope gives a value of $-K_m$. The Eadie-Hoftsee plot will tend to magnify departures from linearity which may not be so apparent in the Lineweaver-Burke transformation.

In general, it is recommended that a K_m value of substrate be included in a one-substrate enzyme assay. Below this value, the signal-to-noise ratio may be too low. Use of concentrations above the K_m will increase the signal-to-noise ratio but also make the assay less sensitive in detecting competitive inhibitors as the enzyme becomes more fully saturated with the substrate. Thus, a balance between signal-to-noise and assay sensitivity needs to be made in developing the enzyme assay.

b) Two-Substrate Reversible Enzyme Reactions

Determination of the K_m of two-substrate enzyme reactions is similar to the method used for one substrate reactions. For the reaction:

$$\mathbf{A + B \underset{}{\overset{E}{\rightleftharpoons}} P + Q}$$

the concentration of B is fixed at a saturating level and the concentration of A is varied to determine its effect on the reaction velocity and the K_m value for A, designated as K_{mA}. Generally a minimum of three different fixed concentrations of B are tested in this analysis. Conversely, K_{mB} is determined experimentally by fixing the concentration of A at a saturating concentration and varying the concentration of A. Two-substrate enzyme reactions generally fall into two different general classes, single- and double-displacement reactions. In single-displacement enzyme reactions, both substrates A and B need to be bound to the enzyme to form a ternary complex. Substrates A and B may associate with the enzyme in either a random or in ordered sequences. In double displacement enzyme reactions, one substrate must bind to the enzyme and one product must be released before the second substrate can bind and dissociate from the enzyme.

Lineweaver-Burke plots (Figure **4**) can be used to distinguish single- and double-displacement mechanisms for two-substrate enzyme reactions. The following plot details the characteristics of a typical ordered two-substrate reaction in which the concentration of A is varied in the presence of three different fixed concentrations of substrate B.

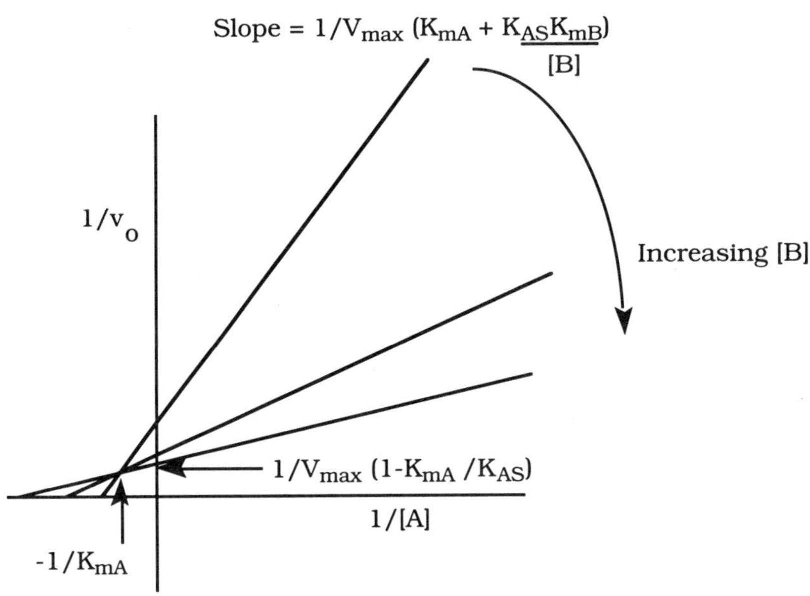

Figure 4: Lineweaver-Burke plot of a ordered two-substrate enzymatic reaction.

The Lineweaver-Burke plot in Figure **5** is representative of a double-displacement two-substrate enzyme reaction. In this case, the concentration of substrate A is varied in the presence of three different fixed concentrations of substrate B.

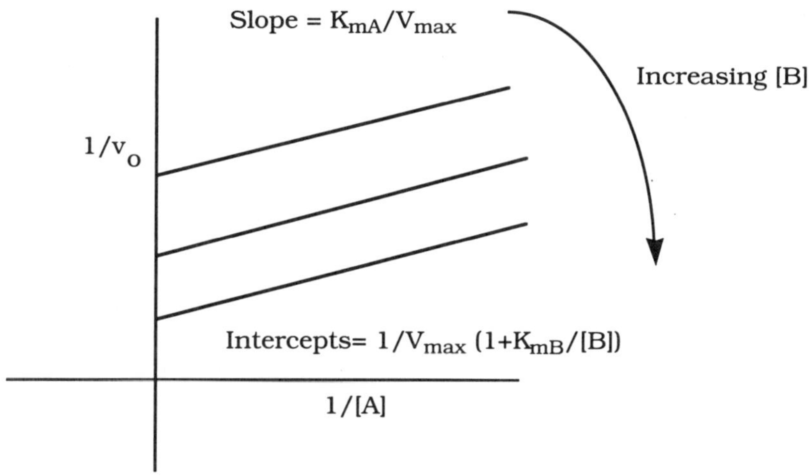

Figure 5: Lineweaver-Burke plot of a double-displacement enzymatic reaction.

Using the above approaches, kinetic parameters for two-substrate enzyme reactions can be experimentally determined. These parameters should be compared to the kinetic information in the literature if it exists. In two-substrate reactions, it should be determined whether the assay should be biased to detect inhibitors of substrate A or B interaction with enzyme. For example, let us consider that a protein kinase assay is to be developed. Protein kinase enzymatic reactions involves two substrates, ATP and the peptide that is to be phosphorylated. If the intent of the screen is to identify compounds that inhibit the binding of the peptide phosphate acceptor and not the inhibitors of ATP binding, the assay should include a saturating amount of ATP and a K_m value of peptide that is phosphorylated. In this way the assay is biased to be more sensitive to picking up competitive inhibitors of only one of the two substrates in the assay.

2.2.2.4 Selection of Cofactor Concentrations

Many enzymes require cofactors for maximal activity. It is not the scope of this section to describe different types of cofactors and their role in catalysis for different enzymes. However, it is imperative that necessary cofactors be identified and included when developing enzyme assays. Once cofactors are identified, it is important to determine what concentrations are needed for optimal enzyme activity. In general, concentrations exceeding saturation should be included in screening assays so that inhibitors of cofactor binding or cofactor depletors are not identified in the screening assay. For example, an assay for the enzyme NO synthase should contain saturating amounts of the required cofactors NADPH, tetrahydrobiopterin, calcium, and calmodulin.

2.2.2.5 Selection of Assay Endpoint

For any enzyme assay, there are usually several different endpoints that could be selected for the assay. These different types of assays would generally include radiometric, colorimetric, and fluorometric assays. Radiometric assays are very sensitive and are amenable to assay automation. However, many radiometric assays require resolution of labeled substrate and product. One exception are radiometric enzyme assays using the SPA technology. Radiometric assays are declining in use in high volume screening assays due to safety considerations and high costs associated with disposal of radioactive waste. Colorimetric assays usually possess lower sensitivity but are also amenable to high throughput screening. One of the paramount problems with screening natural product extracts in colorimetric assays is that many natural product extracts are colored and these can interfere with the endpoint measurement. Fluorometric assays are highly sensitive and are amenable to high throughput screening. Similar to colorimetric assays, natural product extracts may interfere with endpoint reading due to the presence of fluorescent compounds or fluorescence quenching compounds. Fluorescence-based assays appear to be gaining more popularity in screening programs and may also contain sufficient signal-to-noise ratios to permit assay miniturization which is anticipated in the near future. Fluorescence polarization and time-resolved fluorescence assays are being applied to enzyme assays such as protease assays.

2.2.2.6 Selection of Assay pH

Most enzymes have a pH value at which the activity is maximal. The relationship of enzyme activity to pH for most enzymes is a bell-shaped function. In developing an enzyme assay, it is important to determine the optimal pH by simply measuring activity at different pH values in the presence of saturating concentrations of substrate. The shape of the above relationship will also vary with substrate concentration as the K_m of a substrate is typically dependent upon pH.

2.2.2.7 Selection of Assay Temperature

Assay temperature is also an important variable to evaluate when developing an enzyme assay. In general, the rate of an enzyme reaction increases with increasing temperature. The rate for most enzyme reactions doubles with each increase in 10°C. However, at some increasing temperature, enzymes can become denatured or more susceptible to proteolysis, thus limiting incubation temperatures. From a practical standpoint, it is advantageous to perform enzyme assays at ambient room temperature so that incubators are not required. However, for certain low V_{max} enzymatic reactions, higher temperatures may be required so that incubations can be performed for a practical and convenient time period.

2.2.2.8 DMSO Compatibility

Dried natural product extracts are typically resuspended in DMSO before testing in assays. The higher the concentration of DMSO present in assays, the higher the probability that secondary metabolites will remain in solution. It is important in developing an enzyme assay that the tolerance of the assay to different DMSO concentrations be determined.

2.2.2.9 Extract Compatibility

Many natural product screening programs will involve screening of extracts from several different sources including actinomycetes, fungi, marine organisms, and plants. In addition, several different chemical extracts may also be made for any given organism to select for certain types of secondary metabolites. These different types of extracts need to be evaluated in the assay to determine the optimal concentration to include in screening. This is accomplished by screening a relatively large sample of different extracts and determining the final assay concentration which gives a desirable number of active compounds or assay hits. Generally, a hit rate of between 0.1% and 1% is desirable. The compatibility of various types extracts for a given assay is often related to the type of assay endpoint that is used.

2.2.2.10 Inhibitor Profiles

An integral part of developing an enzyme assay for identification of inhibitors is to evaluate several different inhibitors in the assay, if they are available. Both the inhibitor K_i value and the mechanism of inhibition can be determined and compared with the information in the literature. Although enzyme activators may be of interest, the majority of enzyme assays are configured to identify enzyme inhibitors. There are generally three types of reversible enzyme inhibition: competitve, uncompetitve, and noncompetitve. These different mechanisms can be experimentally distinguished by evaluating the effect of increasing fixed concentrations of inhibitor on enzyme V_{max} and K_m values. The following discussion details how the K_i and mechanism of inhibition can be determined experimentally.

2.2.2.11 Competitive Inhibition

The main characteristic of a competitive inhibitor is that it competes with the substrate for binding to the enzyme active site. Competitive inhibition can be decreased by increasing the concentration of substrate. Lineweaver-Burke plots of $1/v_o$ vs $1/[S]$ performed in the presence of increased fixed concentrations of a competitive inhibitor will result in an increase in K_m values without affecting V_{max} values, as shown in Figure **6**.

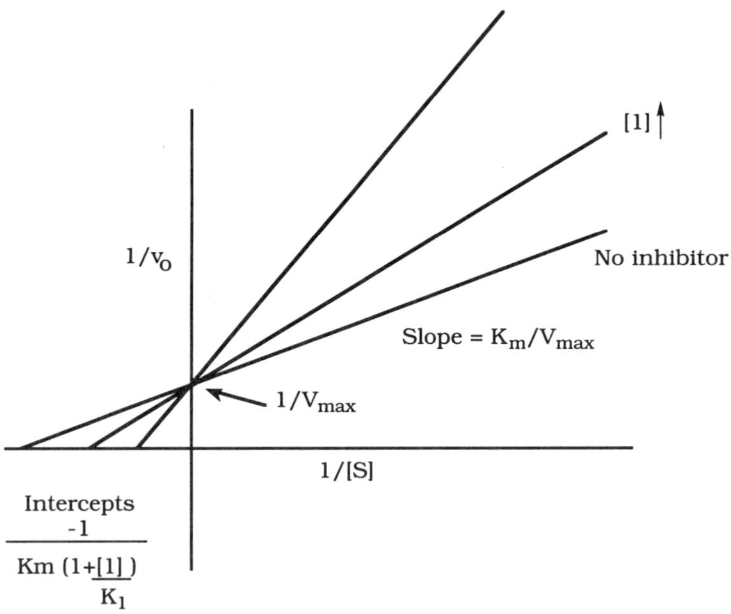

Figure 6: Lineweaver-Burke plot of competitive inhibition.

From the relationship between the structure of a competitive inhibitor and its competition for substrate binding, valuable information concerning the enzyme active site can often be obtained.

2.2.2.12 Uncompetitive Inhibition

During uncompetitive inhibition, the inhibitor does not interact with the free enzyme but interacts with the enzyme-substrate complex to inhibit product formation. Therefore, K_m and V_{max} values will decrease in the presence of increasing inhibitor. Lineweaver-Burke plots of data generated in the presence of increasing concentrations of uncompetitive inhibitor will have the general features shown in Figure **7**.

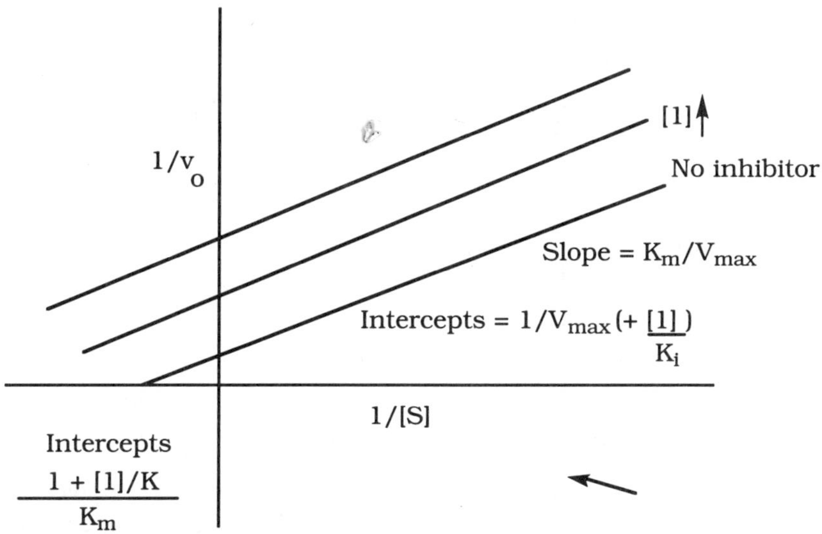

Figure 7: Lineweaver-Burke plot of uncompetitive inhibition.

Uncompetitive inhibition is very rare in a one-substrate enzymatic reaction and common in two-substrate enzyme reactions.

2.2.2.13 Noncompetitive Inhibition

Noncompetitive inhibitors can interact with either the free enzyme or with the enzyme-substrate complex. Noncompetitve inhibitors bind to sites on the enzyme other than the active site and promote conformational changes in the enzyme resulting in a nonfunctional enzyme that does not form a substrate-enzyme complex or that decomposes to yield one product at a typical rate. Noncompetitive inhibition is generally very rare and in many cases is referred to as "mixed inhibition". Lineweaver-Burke plots (Figure **8**) of activity in the presence of increasing concentrations of inhibitor will indicate a decrease in V_{max} and no change in K_m values. Figure 8 shows a charcteristic Lineweaver-Burke plot of noncompetitive inhibition.

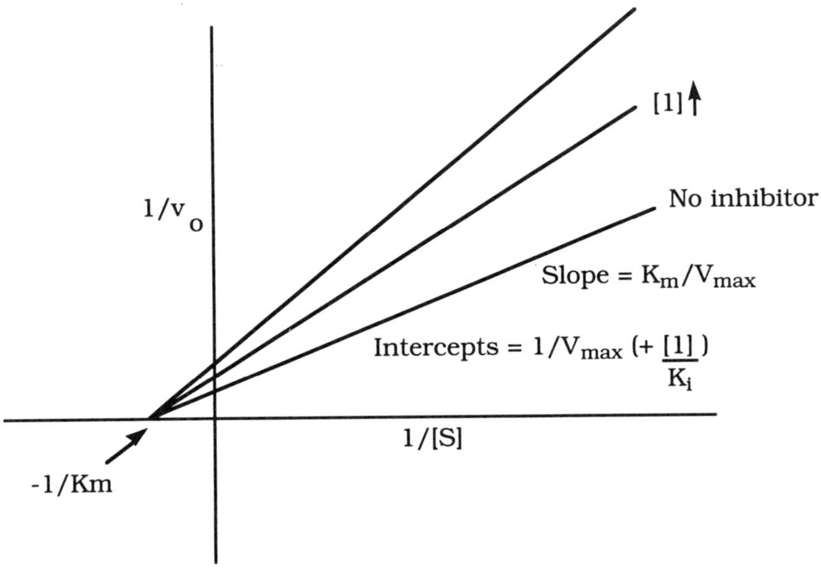

Figure 8: Lineweaver-Burke plot of noncompetitive inhibition.

From a drug discovery standpoint, a noncompetitive inhibitor would not be as desirable as a competitive inhibitor and would be expected to be more nonselective and most likely to inhibit multiple enzymes.

The above relationships also hold for inhibitors of two-substrate enzyme reactions. In this case, the influence of increasing concentrations of inhibitor are evaluated for their influence on the relationships of substrate A and substrate B concentrations on K_m and V_{max} values determined from the Lineweaver-Burke plots.

2.2.3 Development of a High Throughput Enzyme Assay for Natural Product Screening: An Example

To illustrate several of the above points made concerning development of a high throughput enzyme assay for natural product screening, the development of an enzyme assay for farnesyltransferase will be presented.

Farnesyltransferase is an enzyme that catalyzes the addition of a C-15 farnesyl isoprenoid moiety onto specific protein substrates possessing a C-terminal amino acid CAAX box (Casey et al., 1989). Ras is one physiological substrate for farnesyltransferase and this modification has been shown to be essential for Ras associated cellular transformation [Cox *et al.*, 1992; Schafer *et al.*, 1992]. Several forms of human cancer that are resistant to conventional anticancer drugs, such as colon and pancreatic cancers, contain high levels of mutated Ras protein [Lowry *et al.*, 1993]. Due to these observations, it has been postulated that inhibitors of Ras farnesylation may be valuable anticancer agents. Several different peptide-like and natural product inhibitors of farnesyltransferase have been identified in screening programs and some of these have been reported to inhibit transformation of normal cells by mutant Ras proteins [Gibbs et al., 1994].

Farnesyltransferase is a zinc containing cytosolic heterodimeric protein containing $\alpha-$ and $\beta-$subunits which requires both zinc and magnesium ions for activity [Chen et al., 1991; Reiss *et al.*, 1990]. The α-subunit binds farnesylpyrophosphate and is shared with the related isoprenylation enzyme, geranylgeranyltransferase. The β-subunit binds to substrate. It is structurally distinct from the β-subunit of geranylgeranyltransferase and binds to the substrate that becomes isoprenylated. The farnesyltransferase assay is thus a two-substrate enzymatic reaction requiring both farnesylpyrophosphate and protein substrate containing the N-terminal CAAX box. Detailed kinetic studies have indicated that there is a random order of substrate addition to the enzyme (Pompliano et al., 1992). In configuring the assay for drug discovery, a decision needed to be made as to what type of inhibitor the screen should be biased to identify. Competitive inhibitors of farnesylpyrophopshate binding would not be desired because of a potential cross-reactivity of these inhibitors with geranylgeranyl-transferase, resulting in a lack of selectivity and possibly severe adverse effects. A screen designed to identify competitive inhibitors of substrate binding to the β-subunit would be more desirable because the potential for selectivity for only farnsyltransferase would be much higher.

The first consideration for developing an assay was to identify an appropriate source of the farnesyltransferase enzyme. Rat brain was chosen because no suitable human sources could be identified. However, human placenta has been recently suggested to be a good source of human enzyme. The enzyme was partially purified (50 fold) by subjecting 100,000 x g supernatants to ammonium sulfate precipitation and MonoQ and Sephacel S-200 column chromatography. A normal preparation using

50 rat brains provided enough material for approximately 100 assay plates. These preparations were checked for geranylgeranyltransferase activity and none was detected. Studies of enzyme stability indicated that the partially purified enzyme was stable with storage at -80°C for a period of one month. Subsequent to developing this assay, a patent was awarded to the University of Texas Southwestern Medical Center covering the isolation of farnesyltransferase. This technology can be licensed for use in drug discovery.

Several different assays for farnesyltransferase could be considered for developing a screening assay. The most attractive from a high throughput standpoint was a farnesyltransferase scintillation proximity assay kit commercially available from Amersham. The major attraction of this assay was that a separation step normally required to resolve substrates and product in typical radiometric assay methods is not required in the SPA assay. Figure **9** shows a diagramatic representation of the principle and the steps associated with the SPA assay.

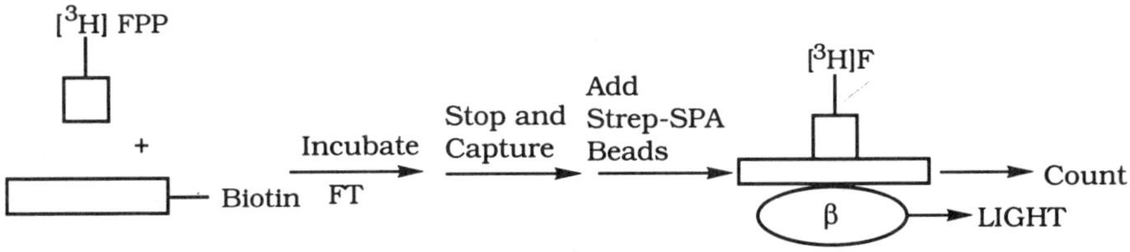

Figure 9: Principle of the Amersham SPA farnesyltransferase assay.

The principle of the SPA assay is that the [^{3}H] farnesyl moiety is incorporated into a biotinylated synthetic peptide corresponding to the C-terminal tail of lamin, a physiological substrate for farnesyltransferase. After incorporation, streptavidin coupled to SPA beads is added. By virtue of the high affinity interaction of streptavidin and biotin, the [^{3}H]-farnesyl moiety incorporated into the lamin peptide is brought into close proximity with the scintillant containing SPA bead. Unincorporated [^{3}H]- farnesyl moiety not in close proximity with the SPA beads, does not contribute to the signal. The actual assay steps involved in the SPA assay are as follows:

1. 50μl of assay buffer, 15μl of 0.5μM biotinyl-lamin peptide, and 15μl of 60μM [^{3}H]-farnesylpyrophos-phate are added to microtiter plates. Negative controls also contain 2 μM CVLS peptide.
3. Test compounds are added in 10 μl aliquots.

5. The enzyme reaction is initiated by the addition of 10 μl of enzyme (1 μg).
7. Plates are incubated for precisely 60 minutes at 37°C.
9. 150 μl of stop buffer containing streptavidin-coated SPA beads is added to each well and the plates are incubated for 30 minutes at room temperature.
11. Plates are then read in a microtiter plate scintillation counter.

This assay is well suited for high volume screening in that only two addition steps are required before endpoint reading. Since the assay is in a micotiter plate format, it is amenable to robotic manipulation and can be completely automated.

Although the assay kit provided by Amersham contains an assay protocol, it is necessary to optimize the assay for screening. The first step in the assay development and validation is to evaluate the relation of enzyme concentration to the assay signal-to-noise ratio. This experiment

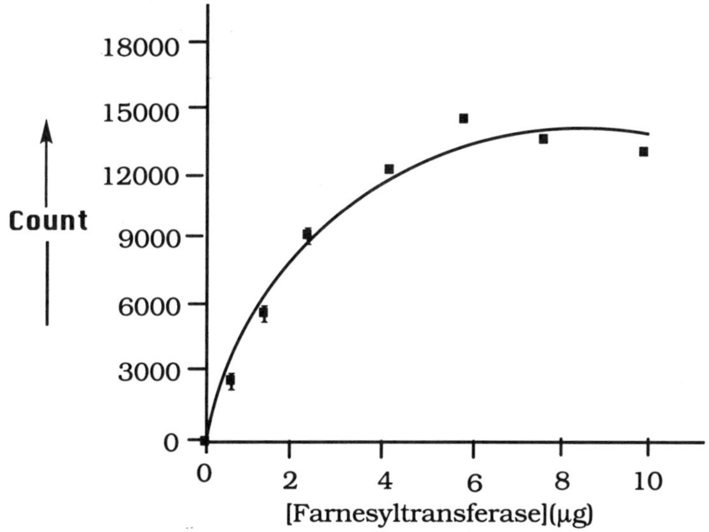

Figure 10: Variable addition of farnesytransferase in the SPA assay. Final concentrations of 9 mM [^{3}H]-FPP, 100 nM biotinyl-lamin peptide, and indicated concentrations of partially purified enzyme were present in the assay and incubations were performed for 60 minutes at room temperature. Data points represent the mean+/-SEM of triplicate determinations at each enzyme concentration. Data were fit to a hyperbolic function using nonlinear least squares curve fitting program, Prism.

was performed in the presence of saturating concentrations of [^{3}H]-farnesylpyrophosphate and a concentration of biotinyl-lamin which was slightly higher than the K_m value for enzyme. Experimental data indicated that the activity was linear up to 2 μg of added protein. At 1 μg, the signal-to-noise was 100:1 and acceptable for the screening assay (Figure **10**).

The next experiment was designed to determine the K_m value for [^{3}H]-farnesylpyrophosphate and was performed in the presence of a saturating concentration of biotinyl-lamin (0.075 μM) and 1 μg of the enzyme. Results of this experiment indicated that [^{3}H]-farnesylpyrophosphate had a K_m value of approximately 2.3 μM (Figure **11**). A decision was made to include 10 μM final concentration of [^{3}H]-farnesylpyrophosphate to ensure that the enzyme was fully saturated with respect to this substrate and more biased in detecting inhibitors of biotinyl-lamin interaction with farnesyltransferase.

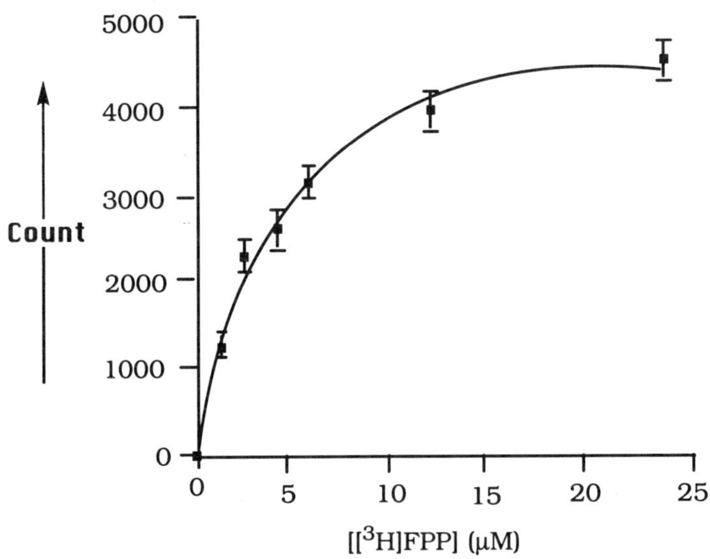

Figure 11: Variable addition of [^{3}H]-farnesylpyrophosphate in the SPA farnesyltransferase assay. Final concentrations of 1mg of enzyme, 100 nM biotinyl-lamin, and indicated concentrations of [^{3}H]-farnesylpyrophosphate were present in the assay; incubations were performed for 60 minutes at room temperature. Data points represent the mean+/-SEM of triplicate determinations made at each [^{3}H]-FPP concentration. Data were fit to a hyperbolic function using the nonlinear least squares curve fitting program, Prism.

The K_m value for biotinyl-lamin was next determined in the presence of a saturating concentration of [^{3}H]- farnesylpyrophosphate and 1 μg of enzyme (Figure **12**). A K_m value of 19 nM was determined. Due to a relatively low signal-to-noise at 20 nM, a decision was made to include 70 nM biotinyl-lamin in the assay to achieve a high signal-to-noise ratio in the range of 100:1.

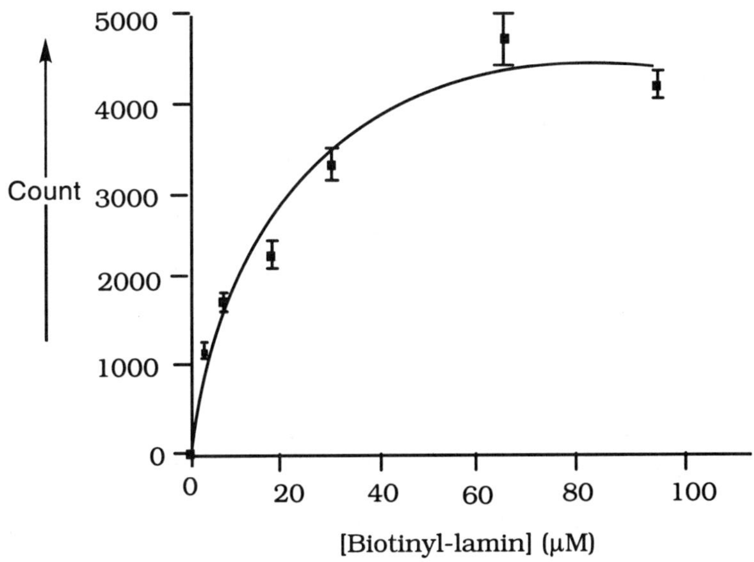

Figure 12: Variable addition of biotinyl-lamin in the SPA farnesyltransferase assay. Final concentrations of 1mg of enzyme, 10 mM [^{3}H]-FPP, and indicated concentrations of biotinyl-lamin were present; incubations were performed for 60 minutes at room temperature. Data points represent the mean+/-SEM of triplicate determinations made at each indicated biotinyl-lamin concentration. Data were fit to a hyperbolic function using the nonlinear least squares curve fitting program, Prism.

A time course evaluation was next performed in the presence of 10 μM [^{3}H]-farnesylpyrophosphate, 70 nM biotinyl-lamin, and 1 μg of enzyme/well (Figure **13**). Results of this evaluation indicated that the % [^{3}H]-farnesylpyrophosphate incorporated increased linearly between 10 and 60 minutes at room temperature. An incubation period of 60 minutes at room temperature was chosen for subsequent assays. At this time period, incubations are under initial rate conditions.

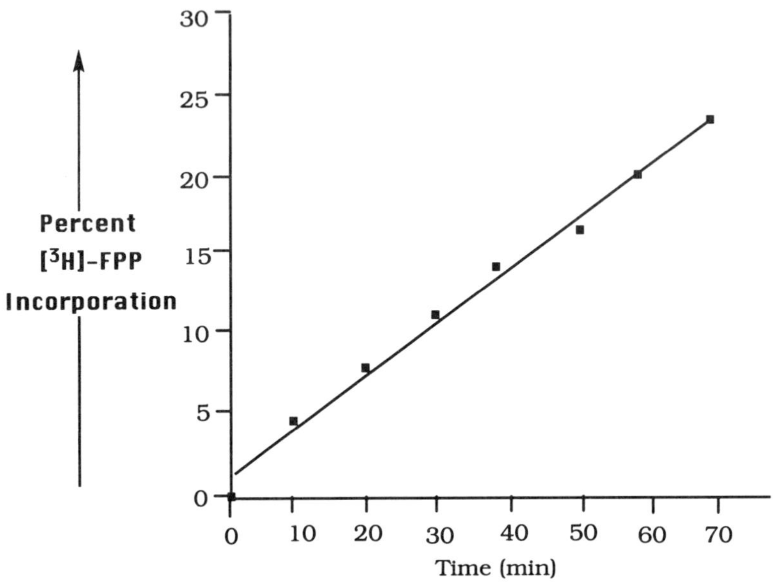

Figure 13: Farnesyltransferase assay time course. The assay contained final concentrations of 1 mg enzyme, 10 mM [^{3}H]-FPP, and 70 nM biotinyl-lamin and incubations were performed for indicated periods at room temperature. Triplicate determinations were made at each time point, the mean was calculated, and the percentage of added [^{3}H]-FPP that was incorporated into substrate is indicated. The line drawn represents linear regression of the data using the nonlinear least squares curve fitting program, Prism.

During development of this assay, no inhibitors were commercially available for evaluation in the assay. A tetrapeptide CVSL, which corresponds to the CAAX box of p21-Harvey Ras, was used to inhibit farnesylation of biotinyl-lamin. Since CVLS is a substrate for farnesyltransferase, it would compete for farnesylation with biotinyl-lamin. This peptide inhibited incorporation of [^{3}H]-farnesylpyrophosphate into biotinyl-lamin with an IC_{50} value of 796 nM (Figure **14**).

The influence of DMSO on farnesyltransferase activity was evaluated next to determine the assay tolerance for DMSO. Activity was unaffected by DMSO concentrations upto 10% and this value of final DMSO concentration was used in the screening assay (data not shown). Several different types of ethyl acetate extracts were evaluated in the assay to establish the final concentration of extract that would result in a hit rate of approximately 0.5%. A final concentration of 0.1X (of the original 10X concentrated extract) for 200 different terrestrial and marine actinomycete extracts and 200 different terrestrial and marine fungal extracts was found to generate the desired assay hit rate. Organic methylene chloride extracts, aqueous ethanol extracts, and aqueous ethanol extracts subjected to polyphenol removal by chromatography prepared from for 100 different plants were also tested in the assay. A final assay concentration of 40 mg/ml for all three different plant extracts gave a hit rate of approximately 0.5%.

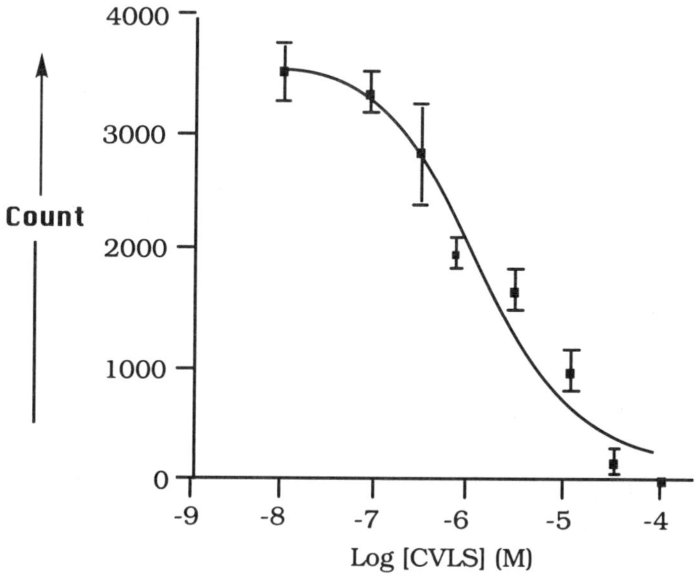

Figure 14: CVLS competition for farnesylation of biotinyl-lamin. The assay contained final concentrations of 1 mg enzyme, 10 mM [^{3}H]-FPP, 70 nM biotinyl-lamin, and indicated concentrations of CVLS; incubations were performed for 60 minutes at room temperature. Data points, representing the mean+/-SEM of triplicate determinations, were made at indicated CVLS concentrations. The curve represents the fit of data to a one-site competition equation using the nonlinear least squares curve fitting program, Prism.

2.2.4 Some Common *In Vitro* Enzyme-Based Bioassays

2.2.4.1 Protease Inhibition Assays

Proteases or proteinases are the proteolytic enzymes which play a vital role in the normal physiological functions of cells e.g. protein maturation, digestion, blood coagulation, control of blood pressure, immune response, etc. A variety of diseases such as cancer, pulmonary emphysema, muscular dystrophy, arthritis, pancreatitis, etc. are associated with the excessive activity of proteases. The role of proteases in diseases therefore provides targets for the possible treatment of a wide range of diseases by protease inhibitors as therapeutic agents from natural sources.

1. Proteases and their specific chromogenic substrates:

	Protease	Substrate
(i)	α-Chymotrypsin	N-Suc-phe-*p*-nitroanilide
(ii)	Trypsin	Bz-DL-Arg-*p*-nitroanilide
(iii)	Elastase	Suc-$(Ala)_3$-*p*-nitroanilide
(iv)	Carboxypeptidase A	Hippuryl-L-phenylalanine
(v)	Papain	Bz-DL-Arg-*p*-nitroanilide

Materials

1. *Tris* (hydroxymethyl) aminomethane
2. Buffer **A** (pH 7.5):
   Tris-HCl (1M)
   NaCl (0.5 M)
3. Buffer **B** (pH 8.6):
   Tris-HCl (0.4 M)
   $MgCl_2$ (5 mM)
4. Buffer **C** (pH 7.5):
   Tris-HCl (50 mM)
   L-Cysteinium chloride (1mM)
   EDTA.Na_2 (2 mM)
5. Lithium chloride (10% w/v)
6. DMSO
7. HCl (5 M)
8. Volumetric flasks
9. Measuring cylinders
10. Beakers
11. Micropipettes
12. pH meter
13. UV/Vis spectrophotometer
14. Deionized water
15. Timer
16. Eppendorf tubes (siliconized)
17 96 well microplates (flat bottom)
18 Microplate reader
19. Test sample (crude extract, pure natural or synthetic compounds)

Principle: Chromogenic substrates of proteases have a specific amino acid sequence linked to a chromophore such as *p*-nitroaniline. The action of a specific protease on its substrate causes the release of the chromophore which is measured as increase in absorbance in a recording spectrophotometer. Therefore, the amount of chromophore liberated is proportional to the activity of the enzyme.

a) Preparation of Solutions:

1) Tris-HCl Buffer

121.14 gm of Tris (hydroxymethyl)-aminomethane are dissolved in deionized water. The pH is adjusted to the required value with HCl (5 M), and the volume is made upto 1 liter with water.

2. Assay buffers for proteases:

Protease	Buffer	Molar Conc. (M)	Buffer pH
1. α-Chymotrypsin	Tris-HCl	0.4	7.5
2. Trypsin	Tris-HCl	0.4	7.5
3. Elastase	Tris-HCl	0.4	8.6
4. Carboxypeptidase A	A	-	7.5
5. Leucine amino-peptidase	B	-	8.6
6. Papain	C	-	7.5

3. Proteases can be dissolved in the following buffers:

Proteases	Buffer	Conc. (mM)	Buffer pH	Enzyme (Units/ml)
1. α-Chymotrypsin	Tris-HCl	50	7.5	9.0
2. Trypsin	Tris-HCl	50	7.5	150
3. Elastase	Tris-HCl	50	8.6	0.6
4. Carboxypep-tidase A	LiCl (10% w/v)		-	0.4
5. Leucine amino-peptidase	B	-	8.6	4.0
6. Papain	C	-	7.5	6.0

4. Substrates solutions can be dissolved in the following buffers:

Substrates (mM)	Buffer	Molar conc. (mM)	Buffer pH	Substrate conc.
1. N-Suc-Phe-*p*-nitroanilide	Tris-HCl	50	7.5	2.6
2. Bz-*DL*-Arg-*p*-nitroanilide	Tris-HCl	50	7.5	1.0
3. Suc (Ala) 3-*p*-nitroanilide	Tris-HCl	50	8.6	1.55
4. Hippuryl-*L*-Phenylalanine	A	-	7.5	1.70
5. *L*-leucine-*p*-nitroanilide	B	-	8.6	17.40

b) Assay Procedure

Assay buffer 400 µl

Enzyme solution 100 µl
Test sample 1 ml
Substrate solution 1.5 ml

1. 500 µl of assay buffer are dispensed in a clean and dry test tube.
2. 500 µl of protease enzyme solution is added.
3. 1.0 ml of test sample is added and the contents are mixed.
4. The assay mixture is incubated at 37 °C for 30 min.
5. Finally 1.0 ml of substrate solution is added and the absorbance is monitored continuously in a recording spectrophotometer for 15 min. at appropriate wavelength (410 nm for all substrates except for hippuryl-L-phenylalanine, 254 nm).

96 Well Microplate Assay Procedure:

1. 50 µl portions of assay buffer are dispensed in each well of a flat bottom 96 well microplate.
2. 50 µl of protease enzyme solution are added.
3. 100 µl of test sample is added and the contents are mixed.
4. The assay mixture is incubated at 37 °C for 30 min. in a microplate reader.

5. 100 μl of substrate solution are dispensed and the absorbance is monitored continuously in the microplate reader for 15 min. with appropriate wavelengths.
(**Note:** The organic solvent used to dissolve the test sample may affect the enzyme activity).

Controls should be run to measure the change in enzyme activity. Positive controls should also be run with appropriate standard inhibitors. Conditions of the assays should be chosen such that the maximum amount of test sample should be added to each assay with a minimum amount of organic solvent. Although many organic solvents can be used but DMSO is one of the most convenient as it is water miscible and directly compatible with these assay systems+.

Calculation of Percentage Inhibition:

Percentage inhibition in enzyme activity after incubation can be calculated as follows:

$$\% \text{ Inhibition} = \frac{\text{Absorbance (control)} - \text{Absorbance (test)}}{\text{Absorbance (control)}} \times 100$$

2.2.4.2 Tyrosinase Inhibition Assay

Tyrosinase is an enzyme responsible for the synthesis of dermal melanin pigment from L-tyrosine and *L*-DOPA (dihydroxyphenylalanine) within the melanocytes on the melanosomes. It has several functions including hydroxylation of *L*-tyrosine and oxidation of *L*-DOPA to dopaquinone and subsequent autopolymerization to melanin. The over-production of melanin has been associated with the condition of hyperpigmentation of skin such as melasma and ephelides. Therefore the inhibition of tyrosinase enzyme is reflective of controlling hyperpigmentation and associated conditions* .

Materials

1. Phosphate buffer (0.1 M, pH 6.8)
2. *L*-3,4,-Dihydroxyphenylalanine (*L*-DOPA)
3. *L*-Tyrosine
4. Mushroom tyrosinase enzyme
5. DMSO
6. Volumetric flasks

+ For further details see ref. Bergammeyer, 1988 and Barrett *et al.*, 1982

* For further details see Hearing, 1987, Pomerantz, 1963 and Yagi *et al.*, 1987

7. Measuring cylinders
8. Beakers
9. Micropipettes
10. pH meter
11. Timer
12. Water bath
13. Deionized water
14. UV/VIS spectrophotometer
15. Eppendorf tubes (siliconized)
16 96 well microplates (flat bottom)
17 Microplate reader
18. Test sample (crude extract, pure natural product, or synthetic compound)

Tyrosinase first hydroxylates *L*-tyrosine to *L*-DOPA and then oxidizes *L*-DOPA to dopaquinone which is subsequently converted to dopachrome. The activity of tyrosinase enzyme is proportional to the amount of dopachrome liberated that is measured at 475 nm in a spectrophotometer.

1) Preparation of Solutions

(a) Phosphate Buffer (0.1 M, pH 6.8)

(I) **Solution-A (0.2 M $Na_2HPO_4 .2H_2O$)**
35.6 gm. of $Na_2HPO_4 .2H_2O$ is dissolved in water and the volume is made up to 1 liter with water.

(II) **Solution-B (0.2 M, $NaH_2PO_4 .2H_2O$)**
31.2 gm. of $NaH_2PO_4 .2H_2O$ is dissolved in water and the volume is made up to 1 liter with water.
51 ml of solution-B are mixed with 49 ml of solution-A and diluted to a total of 200 ml with water.

(b) Preparation of Enzyme and Substrate Solutions

Enzyme	**Buffer**	**Conc. (M)**	**Buffer pH**	**Enz. conc. (Units/ml)**
Mushroom tyrosinase	Phosphate	0.1	6.8	60
Substrates:				**Substrate conc. (mM)**
(i) L-DOPA	Phosphate	0.1	6.8	2.55
(ii) L-Tyrosine	Phosphate	0.1	6.8	1.70

2) Assay Procedure

Two types of substrates can be used for this assay:

(i) L-DOPA and (ii) L-Tyrosine

(1) 1.0 ml of phosphate buffer is dispensed in a test tube.

(2) 500 μl of mushroom tyrosinase enzyme solution is added.

(3) 500 μl of test sample is added, mixed and incubated at 25 °C for 10 min.

(4) 1.0 ml substrate solution is finally added.

(5) The absorbance is monitored continuously in a recording spectrophotometer at 475 nm. for 20 min.

3) 96 Well Microplate Assay Format:

(1) 100 μl of phosphate buffer are dispensed in each well of a 96 well microplate.

(2) 50 μl of mushroom tyrosinase enzyme solution are added.

(3) 50 μl of test sample are added, mixed and incubated at 25 °C for 10 min.

(4) 100 μl substrate solution are finally added.

(5) The absorbance is monitored continuously in a microplate reader at 475 nm. for 20 min.

4) Calculation of Percentage Inhibition:

The percentage inhibition in enzyme activity can be calculated as follows:

$$\% \text{ Inhibition} = \frac{\text{Absorbance (control)} - \text{Absorbance (test)}}{\text{Absorbance (Control)}} \times 100$$

2.2.4.3 Hyaluronidase Inhibition Assay

Hyaluronidase, a mucopolysaccharide hydrolyzing enzyme, has been shown to involve in various diseases. The enzyme is functionally related to vascular permeability and inflammatory reactions. Since the enzyme exists in an inactive form, its *in vivo* activation by metal ions including calcium ions is related to the degranulation of mast cells, causing release of mediators that cause allergy including inflammation.

Hyaluronate, a mucopolysaccaharide, consisting of repeating subunits of D-glucuronic acid and *N*-acetyl-D-glucosamine, is a viscous lubricating agent present in synovial fluid in joints. In rheumatoid arthritis, its excessive degradation by hyaluronidase may lead to decrease in amount and molecular weight of hyaluronate thus producing arthritic symptoms. The search for inhibitors of hyaluronidase enzyme may lead to the isolation of new potent anti-allergic and anti-inflammatory drugs.+

Materials

1. Acetate Buffer (0.1 M, pH 3.5)
   Sodium acetate (0.1 M)
   Glacial acetic acid (0.1 M)
2. Hyaluronidase enzyme (from bovine testes)
3. Potassium hyaluronate
4. Calcium chloride (2.5 mM)
5. Sodium hydroxide (0.4 N)
6. Potassium tetraborate (0.8 N, pH 9.1 adjusted with 5 M KOH)
7. *p*-Dimethylaminobenzaldehyde
8. Screw cap test tubes (Pyrex)
9. Micropipettes
10. Water bath
11. HCl (10 N)
12. Timer
13. UV/VIS spectrophotometer
14. pH meter
15. Volumetric flasks
16. Beakers
17. Measuring cylinders
18. Deionized water
19. Test sample (crude extract, pure natural product or synthetic compound)

Principle:

Hyaluronidase enzyme hydrolytically cleaves the β (1-4) bond in hyaluronic acid liberating the product with a terminal N-acetyl-D-glucosamine moiety that reacts with alkali to form a glucoxazoline intermediate compound. This intermediate reacts with *p*-dimethylaminobenzaldehyde to produce a colored product which is measured at 585 nm.

+ For further details of this assay see Reissig *et al.*, 1955; Kakegawa *et al.*, 1985

1) Preparation of Buffers and Solutions:

a) *Acetate Buffer (0.1 M, pH 3.5)*

(I) Section A:
16.4 gm of sodium acetate is dissolved in water and the volume is made up to 1 liter with water.

(II) Section B:
11.55 ml of glacial acetic acid is mixed with water and the volume is made up to 1 liter with water.
46.3 ml of solution-B are mixed with 3.7 ml of solution A and diluted to a total of 100 ml with water.

b) *Solvent for p-dimethylaminobenzaldehyde:*

12.5 ml of HCl (10 N) is mixed with glacial acetic acid and the volume is made upto 100 ml with the same acid.

c) *p-Dimethylaminobenzaldehyde (67 mM)*

Stock solution:
p-DMAB (10 gm) is dissolved in *p*-DMAB-solvent and the volume is made up to 100 ml with the same solvent.
Shortly before the use, 1 ml of the stock solution is diluted to 10 ml with glacial acetic acid.

d) *Preparation of Enzyme and Substrate Solutions:*

Enzyme:	**Buffer**	**Molar Conc. (M)**	**Buffer pH**	**Enzyme Conc. (N.F.Units/ml)**
Hyaluronidase	Acetate	0.1	3.5	350
Substrate:				**Substrate Conc. (mg/ml)**
Potassium hyaluronate	Acetate	0.1	3.5	1.2

2) Assay Procedure

a) *Activation of Hyaluronidase Enzyme*

(I) 400 μl of hyaluronidase enzyme solution (350 N.F. units/ml of acetate buffer) is taken in a screw cap test tube.

(II) 100 µl of $CaCl_2$ solution (2.5 mM in acetate buffer) is added to this test tube and incubated at 37 °C for 20 min. to activate the enzyme.

b) Inhibition of Activated Hyaluronidase Enzyme

(I) 100 µl of test sample or vehicle is added to the activated enzyme.
(II) This mixture is incubated at 37 °C for 20 min. in a water bath.
(III) 500 µl of potassium hyaluronate (1.2 mg/ml of acetate buffer) is added and incubation is carried out again at 37 °C for 20 min.
(IV) The enzyme reaction is stopped by adding 100 µl of NaOH (0.4 N) and 100 µl of potassium tetraborate (0.8 N, pH 9.1).
(V) The mixture is heated in a boiling water bath for exactly 3 min. and the tubes are cooled water tap water.
(VI) 3 ml of *p*-DMAB solution (67 mM) is added, mixed, and incubated at 37 °C for 20 min. for color development.
(VII) The absorbance of the mixture is measured at 585 nm in a spectrophotometer against blank.

c) Inhibition of the Activation of Inactive Hyaluronidase Enzyme

(I) 400 µl of hyaluronidase enzyme solution (350 N.F. Units/ml of acetate buffer) is taken in a screw cap test tube.
(II) 100 µl of test sample or vehicle is added, mixed and incubated at 37 °C for 20 min.
(III) 100 µl of $CaCl_2$ (2.5 mM in acetate buffer) is added and incubated at 37 °C for 20 min.
(IV) The procedure described from step #3 onward in protocol B is repeated.

Calculation of Percentage Inhibition

The percentage inhibition in enzyme activity can be calculated as follows:

$$\% \text{ Inhibition} = \frac{\text{Absorbance (control)} - \text{Absorbance (test)}}{\text{Absorbance (control)}} \times 100$$

2.2.4.4 *Acetylcholinesterase Inhibition Assay*

Acetylcholinesterase (acetylcholine acylhydrolase, EC 3.1.1.7) plays an important role in the central and peripheral nervous systems, along with the acetylcholine receptor, in the transmission of action potential across nerve-

nerve and neuromuscular synapses. The enzyme's physiological task is the hydrolytic destruction of the cationic neurotransmitter, acetylcholine.

$$\underset{\textit{Acetylcholine}}{(CH_3)_3\overset{\oplus}{N}CH_2CH_2O\text{-}COCH_3} + H_2O \xrightarrow{\text{Acetylcholinesterase}} \underset{\textit{Acetic acid}}{CH_3COOH} + \underset{\textit{Choline}}{(CH_3)_3\overset{\oplus}{N}CH_2CH_2OH}$$

Because of the pivotal role that acetylcholinesterase (AChE) plays in the nervous system, it has long been an attractive target for the rational design and discovery of mechanism-based inhibitors. Some inhibitors of acetylcholinesterase are known to be useful for the treatment of Alzheimer's disease, senile dementia, ataxia and for improving the long-term memory processes by enhancing cholinergic activity.*

2.2.4.4.1 Spectrophotometric Assay

The principle involves the measurements of the rate of production of thiocholine, as acetylthiocholine is hydrolyzed by acetylcholinesterase. Hydrolysis of acetylthiocholine is accompanied by a continuous reaction between the thiocholine liberated and DTNB (dithiobisnitrobenzoic acid) which produces the yellow anion of 5-thio-2-nitrobenzoic acid.

a) $$\underset{\text{Acetylthiocholine}}{(CH_3)_3\overset{\oplus}{N}CH_2CH_2SCOCH_3} + H_2O \xrightarrow{\text{AChE}} CH_3COOH \text{ Acetic acid} + \underset{\text{Thiocholine}}{(CH_3)_3\overset{\oplus}{N}CH_2CH_2SH}$$

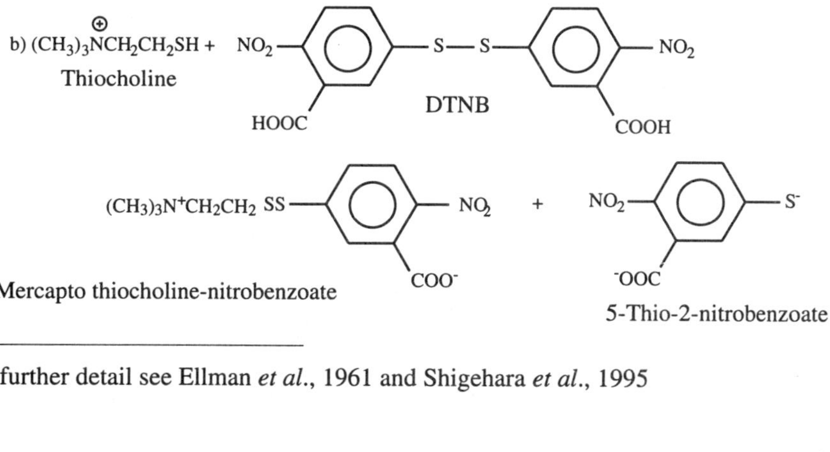

---

* For further detail see Ellman *et al.*, 1961 and Shigehara *et al.*, 1995

The rate of anion production is measured from the absorbance at 412 nm.

Materials

1. Phosphate buffer-1 (0.1 M, pH 8.0)
2. Phosphate buffer-2 (0.1 M, pH 7.0)
3. Buffered Ellman's reagent, DTNB
4. Acetylthiocholine iodide, 75 mmol/l
5. Acetylcholinesterase (AChE)
6. Micropipettes
7. Glass pipettes
8. Stop watch
9. Water bath
10. UV/VIS spectrophotometer
11. Test sample (crude extract, pure natural product, or synthetic compound)

1) Preparation of Reagents

a) Phosphate buffer-1 (0.1 M) (For enzyme and test)

(I) Dissolve 15.6 g of $Na_2HPO_4.2H_2O$ in 750 ml water.
(II) Check the pH at 25°C.
(III) Adjust to 8.0 by adding NaOH solution (100 mmol/1).
(IV) Make volume up to 1 litre by adding distd. water.
(V) Store the buffer solution-1 at 4°C.

This buffer solution is stable as long as no microbial contamination occurs.

b) Phosphate buffer-2 (0.1 M) (For Ellman's reagent, DTNB)

(I) Dissolve 15.6 g of $Na_2HPO_4.2H_2O$ in 750 ml water.
(II) Check the pH at 25°C.
(III) Adjust to pH 7.0 by adding NaOH solution (100 mmol/1).
(IV) Make volume up to 1 litre by adding distd. water.
(V) Store the buffer solution-2 at 4°C

This buffer solution is stable as long as no microbial contamination occurs.

c) Buffered Ellman's Reagent, (DTNB, 0.1M/1; $NaHCO_3$,17.85 mmol/1)

(I) Dissolve 39.6 mg DTNB in 10 ml phosphate buffer-2 solution.

(II) Add 15 mg $NaHCO_3$.
(III) Store in dark bottle at 4 °C.
This solution is stable for 4 weeks if stored in dark bottles.

d) Acetylthiocholine iodide (75 mmol/1) (Substrate)

(I) Dissolve 108.35 mg acetylthiocholine iodide in 5 ml. of water.
(II) Store at 4 °C.
This solution should not be kept for more than 7 days.

e) Acetylcholinesterase (Enzyme)

(I) The enzyme solution is prepared by dissolving the enzyme in phosphate buffer-1 so that the concentration of the enzyme in the reaction mixture is about 0.0025 U/ml.
(II) Keep in iced water bath at 5 °C.

f) Test sample solution

The test sample should be dissolved in the proper solvent (preferably water) with desired concentration, but for water insoluble compounds the effects of other solvents on the enzyme activity should be checked prior to the experiment; the controls should recieve the same volume of the solvent.

The following steps are involved in the acetylcholinestrase inhibition assay:

1. Take 2.81 ml of phosphate buffer-1.
2. Add 30 μl of test sample solution.
3. Add 30 μl of enzyme stock solution.
4. Add 100 μl of DTNB stock solution to this.
5. Incubate for 5-10 min. at 25 °C.
6. Add 30 μl of substrate stock solution.

2) Calculation of Percentage Inhibition

The percentage inhibition in enzyme activity can be calculated as follows:

$$\% \text{ inhibition} = \frac{\text{Absorbance (control)} - \text{Absorbance (test)}}{\text{Absorbance (control)}} \times 100$$

2.2.4.5 β-Lactamase Inhibition Assay

The enzyme β-lactamase is produced by bacteria as a part of a resistance mechanism against β-lactam-containing antibiotics. A large number of β-lactamases (EC 3.5.2.6) have already been identified. β-Lactamases (often called pencillinases, cephalosporinases or carbapenemases, depending on substrate specificity) are plasmid or chromosomally encoded bacterial enzymes which inactivate β-lactam antibiotics. The enzyme efficiently hydrolyzes the amide bond of the β-lactam moiety, thereby yielding products devoid of antibiotic activity.

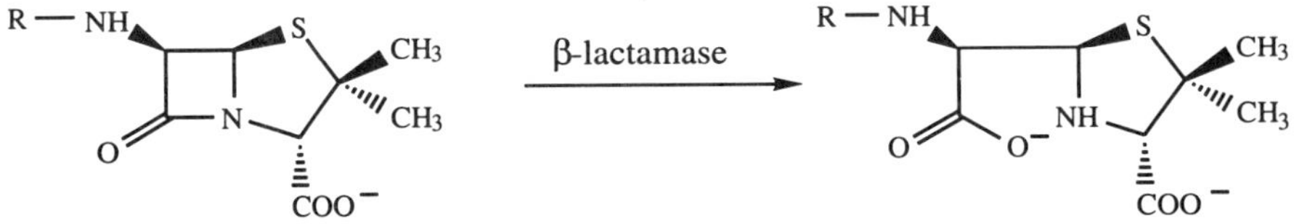

The clinical importance of β–lactamase inhibitors is reflected from the fact that most commonly used antimicrobial agents are β-lactam-containing antibiotics like pencillins, cephalosporins, carbapenems and monobactams. Because of the safety profiles and proven clinical efficacy, these antibiotics are extensively used for the treatment of many infectious diseases. The search for new β-lactamase inhibitors therefore has a tremendous clinical and commercial potential.*

2.2.4.5.1 Spectrophotometric Assay

The measurement of the amount of antibiotics such as nitrocefin hydrolyzed per unit time can be followed spectrophotometrically by the increase in the absorbance at 495 nm due to hydrolysis of the highly conjugated cephalosporin.

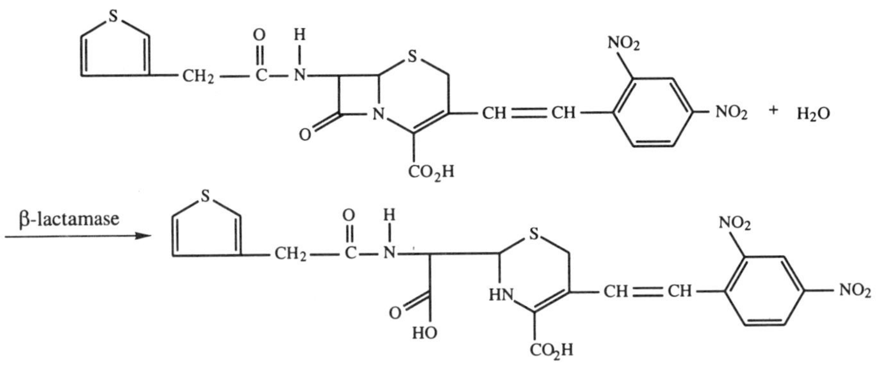

* For further detail see O'Callaghan *et al.*, 1972

Although other pencillins or cephalosporins may be used as substrates in this assay, nitrocefin has the widest spectrum of susceptibility and sensitivity among the commercialy available β-lactam containing antibiotics (for other substrates the UV wavelength range depends upon the nature of the substrate).

Materials

1. Phosphate buffer (100 mmol/l, pH 7.0)
2. Nitrocefin
3. β-Lactamase enzyme
4. DMSO
5. pH paper
6. Distd. water
7. Volumetric flasks
8. Dark glass reagent bottles
9. Micropipettes
10. Stop watch
11. UV/VIS spectrophotometer
12. Test sample (crude extract, pure natural compound, or synthetic compound)

1) Preparation of Reagents

a) Phosphate buffer (100 mmol/l, pH 7.0)

(I) Dissolve 13.4 g $Na_2HPO_4.7H_2O$ in 500 ml. distd. water.
(II) Dissolve 6.85 g of KH_2PO_4 in 500 ml. distd. water.
(III) Mix 80 ml disodium hydrogen phosphate solution with 45 ml dihydrogen potassium phosphate solution.
(IV) Check the pH and adjust to pH 7.0 if necessary with the addition of appropriate phosphate solution. Although an optimum pH of 7.0 has been reported for many of the enzymes, the assay can be performd between pH 6.0 and 8.0 in phosphate buffer (100-200 mmol/l).
(V) Store the solution between 0-4°C. This solution should be prepared freshly at least once a weak.

b) Nitrocefin (0.19 mmol/l)* (Substrate)

(I) Take 5.0 ± 0.2 mg nitrocefin into a 50 ml volumetric flask.

---

* β-lactamase from *Staphylococcus aureus* generally sticks on glass surfaces, therefore, disposable cuvettes should be used for assays of enzymes or glass cuvettes should be rinsed properly with ethanol before adding new samples to prevent carry-over of the active enzyme.

(II) Dissolve it in 0.5 ml DMSO (dimethyl sulfoxide).
(III) Add phosphate buffer (100 mmol/l, pH 7.0) to give a final volume of 50 ml.
(IV) Store in dark bottle at 4°C (if protected from light this solution maintains greater than 90% stability after storage for 2 weeks)

c) *β-Lactamase* Enzyme

Prepare the enzyme solution so that the enzyme concentration will be 550 munit/ml.

d) *Test Sample Solution*

Test sample solution should be prepared in water in the desired concentrations. In case of insoluble compounds, the effect of solvent on the enzyme activity should be checked prior to the experiment and the controls should recieve the same volume of the solvent.

The following steps are involved in the β-lactamase inhibition assays:
1- Preincubate the test sample with the enzyme for 5-10 min.
2- Start reaction by adding 1.0 ml of nitrocefin solution.**
3- Mix well using a plastic mixing spoon.
4- Read absorbance at 495 nm, and start stopwatch. Repeat readings exactly every 15 sec. or monitor on a recorder.

2) Calculation of Percentage Inhibition

The percentage inhibition in enzyme activity can be calculated as follows:

$$\% \text{ inhibition} = \frac{\text{Absorbance (control)} - \text{Absorbance (test)}}{\text{Absorbance (control)}} \times 100$$

2.2.4.6 5-Lipoxygenase Inhibition Assay

The mammalian enzyme 5-lipoxygenase plays an important role in the conversion of arachidonic acid to a number of lipoxygenase derivatives, including 5-HPETE, LTA_4, LTB_4 and peptidoleukotrienes LTC_4, LTD_4 and LTE_4 (Scheme-3).

** Before beginning the assay, the nitrocefin substrate solution should be kept at 30 °C.

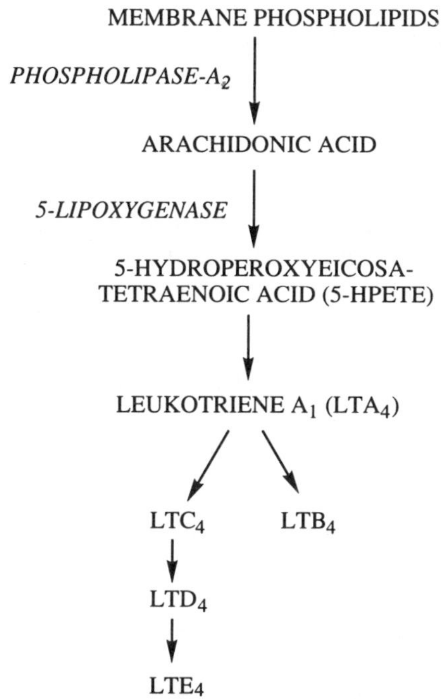

Scheme-3:

The leukotrienes have been identified as a component of SRS-A (slow-reacting substance of anaphylaxis). Since these mediators have been demonstrated to possess potent chemotactic, bronchoconstrictor, and vascular leakage properties they have also been implicated as important mediators of various allergic diseases including asthma. Various lipoxygenase products (LTB_4) also exhibit proinflammatory properties both *in vitro* and *in vivo.* Thus inhibition of 5-lipoxygenase is currently a subject of intense research targeted towards the discovery of novel anti-allergic and anti-inflammatory agents.+

2.2.4.6.1 Spectrophotometric Assay

Lipoxygenases catalyse the reaction of polyenoic fatty acids, containing atleast 1,4-*cis*, *cis*- pentadiene system, with dioxygen forming a 1-hydroperoxy-2-*trans*, 4-*cis* derivative.

+ For further detail see Daniel *et al.*, 1991 and Tanred *et al.*, 1987.

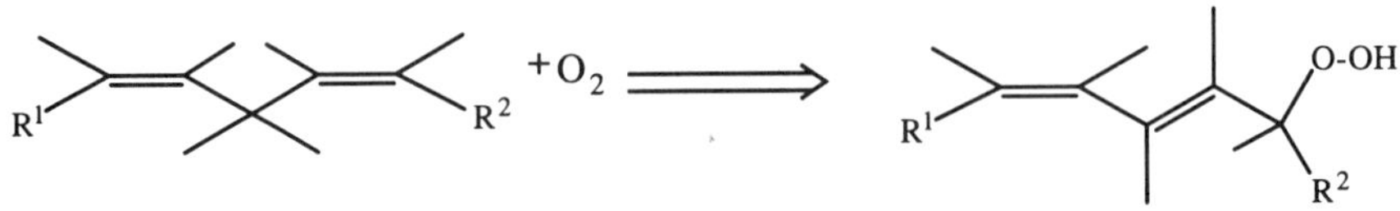

1,4-*cis, cis* -pentadiene 1-hydroperoxy-2- *trans*,4-*cis* -pentadiene

5-Lipoxygenase inhibition activity is measured by the increase in absorbance at 234 nm following incubation of the enzyme with the substrate.

Materials

1. Phosphate buffer (0.1 M, pH 7.4, containing 4% sodium cholate of analytical grade).
2. Substrate mixture (linoleic acid or arachidonic acid).
3. 5-Lipoxygenase
4. Micropipettes
5. Glass pipettes
6. Stop watch
7. UV/VIS spectrophotometer
8. Test (crude extract, pure natural product or synthetic compound).

1) Preparation of Reagents

a) Phosphate buffer (0.1 M, pH 7.4)

(I) Dissolve 15.6g of $Na_2HPO_4.2H_2O$ in 750 ml water.
(II) Check pH at 25 °C.
(III) Adjust to pH 7.4 by adding NaOH solution (100 mmol/l).
(IV) Add 4% sodium cholate.
(V) Further adjust pH if necessary.
(VI) Make volume upto 1 litre by adding dist. water

If stored at 4°C, this solution is stable as long as no microbial contamination occurs.

b) Substrate Mixture ***

(I) Mix 0.2 ml of pure linoleic acid (or arachidonic acid) with 1 ml of freshly distilled methanol.

---

* Keep strict order of additions.

** Do not store the substrate mixture overnight.

(II) Store this stock solution under nitrogen atmosphere at -20°C.
(III) Mix 20 ml of the stock solution with 20 ml of 0.54 M KOH or 0.54 M NH_4OH.
(IV) Add this freshly prepared mixture to 2 ml of cold 0.1 M phosphate buffer.

c) Test Sample

(I) Dissolve test sample in methanol, ethanol, 2-methoxy ethanol (methyl glycol) or DMSO as 0.1 M solution.
(II) Add the test sample solution in the proper dilution in the same solvent in such a volume that the final concentration of the solvent does not exceed 2% of final volume.
(III) Add the same amount of solvent to identically handled control samples.

d) Enzyme Solution

(I) Lipoxygenase enzyme solution is prepared so that the enzyme conc. in the reaction mixture is adjusted to give rates of 0.05 absorbance per min.

5-Lipoxygenase inhibition assay involves following steps:

1. Put 0.95 ml of 0.1 M phosphate buffer, pH 7.4, in precooled quartz cuvette.
2. Add the enzyme solution.
3. Add about 50 µl of the solution of test compound.
4. Incubate for 5-10 min. at 20 °C.
5. Start the reaction by adding 50 µl of substrate mixture.
6. Mix gently and record the absorbance at 234 nm for 2-3 min.

2) Calculation of Percentage Inhibition

The percentage inhibition in enzyme activity can be calculated as follows:

$$\% \text{ inhibition} = \frac{\text{Absorbance (control)} - \text{Absorbance (test)}}{\text{Absorbance (control)}} \times 100$$

2.3 CELL-BASED RECEPTOR FUNCTIONAL ASSAYS

2.3.1 Introduction

2.3.1.1 Utility of Receptor Functional Assays

Endogenous ligand or synthetic agonist binding to specific cytokine, growth factor, hormone and neurotransmitter receptors is the first step in a complex signal transduction pathway which ultimately leads to the generation of defined physiological responses. Signaling pathways typically involve several different signal transduction molecules and are generally shared by structurally related receptors. In contrast to agonist binding, antagonist binding to receptors does not activate the receptor signal transduction pathway and, therefore, antagonists do not cause a cellular response. However, an antagonist will block the response of the endogenous agonist ligand. Although radioligand binding assay provides information concerning ligand binding to a receptor, this type of assay does not provide information concerning whether the receptor ligand is an agonist or antagonist. Assays are available which involve measurement of some aspects of receptor-mediated activation of the receptor signal transduction pathway and thus can serve as functional assays that can be used to distinguish receptor agonists and antagonists. Functional receptor assays are rapidly replacing radioligand binding assays as these assays become cheaper and simpler to perform. One of the primary reasons that functional receptor assays are becoming more popular as primary screening assays is that they have been simplified and are configured into a 96-well microtiter plate format for high throughput screening of a large numbers of extracts. The following section details several functional receptor assays that may be used as primary screens for screening combinatorial chemistry reaction products and natural product extracts.

2.3.2 Functional Assays for G-protein Coupled Receptors

2.3.2.1 Gs and Gi Coupled Receptor Signal Transduction Pathways

Agonist occupation of several different receptors belonging to the seven transmembrane receptor (7-TM) or G-protein coupled receptor (GPCR) superfamily results in either increasing or decreasing concentrations of the important intracellular second messenger, cAMP (Tang *et al*, 1993; Taussig *et al*, 1995) (Table **3**). This receptor-mediated response is due to a G-protein-mediated stimulation or inhibition of the enzyme adenylyl cyclase. Increases in cAMP in turn lead to activation of the specific serine/threonine protein kinase, cAMP-dependent protein kinase or protein kinase A (PKA). The majority of physiological responses mediated by adenylyl cyclase coupled receptors involve increases or decreases in PKA-mediated phosphorylation of specific protein substrates. The cellular levels of cAMP formed by adenylyl cyclase are tightly controlled by the action of phosphodiesterase isozymes which catalyze the conversion of cAMP to AMP (Butcher *et al.*, 1962; Beavo, 1995) (see Figure 15.). The mechanism by which hormones or neurotransmitters increase cellular cAMP levels has been well characterized and is described below:

Agonist binding to its perspective receptor leads to coupling of the agonist-receptor complex with a specific heterotrimeric guanine nucleotide binding protein (G-protein), Gs. The ternary complex formed between agonist-receptor and Gs constitutes the high affinity agonist binding conformation of the receptor whereas receptor not associated with Gs constitutes the low affinity agonist binding conformation (DeLean et al., 1980; Wreggett *et al.*, 1984; Elhert, 1985; Neubig et al., 1985; Neubig et al., 1988). The formation of the ternary complex results in the release of bound GDP from inactive Gs, the subsequent binding of GTP leading to Gs activation (Cassel *et al*, 1979). Gs activation is postulated to involve the dissociation of Gs into its constitutive α_s-GTP and $\beta\gamma$ subunits which act to activate adenylyl cyclase. Activation of adenylyl cyclase by Gs-protein is terminated by the hydrolysis of GTP to GDP (referred to as the GTPase reaction) followed by reassociation of Gs subunits. Screening assays designed to search for agonists of Gs-coupled receptors involve testing compounds for stimulation of basal adenylyl cyclase activity. Screens designed to identify antagonists of Gs-coupled receptors involve screening for compounds that decrease agonist-mediated adenylyl cyclase stimulation.

Receptor-mediated inhibition of adenylyl cyclase shares many common features with stimulation. Agonist binding to receptor promotes coupling with the specific G-protein, Gi (Jakobs et al., 1981; Bokoch et al., 1983), leading to release of GDP, binding of GTP, Gi subunit dissociation, and inhibition of adenylyl cyclase by Gi subunits. As with Gs, Gi undergoes the GTPase reaction resulting in reformation of inactive trimeric GDP bound Gi. Screens designed to identify agonists of Gi coupled receptors usually are designed to identify compounds that inhibit forskolin-stimulated adenylyl cyclase activity. Identification of Gi receptor antagonists is accomplished by screening compounds for their ability to reduce agonist-mediated inhibition of forskolin-stimulated adenylyl cyclase activity.

Table-3: Gs and Gi Coupled Receptors

Receptor	Coupling	Receptor	Coupling	Receptor	Coupling
Adenosine A_1	Gi	Melatonin ML_{1A}	Gi	Prostanoid EP_4	Gs
Adenosine A_{2A}	Gs	Melatonin ML_{1B}	Gi	Purinoceptor P_{2T}	Gi
Adenosine A_{2B}	Gs	Metabotropic $mGlu_2$	Gi	Secretin	Gs
Adenosine A_3	Gi	Metabotropic $mGlu_3$	Gi	Serotonin $5HT_{1A}$	Gi
Adrenergic a_{2A}	Gi	Metabotropic $mGlu_4$	Gi	Serotonin $5HT_{1B}$	Gi
Adrenergic a_{2B}	Gi	Metabotropic $mGlu_6$	Gi	Serotonin $5HT_{1D}$	Gi
Adrenergic a_{2C}	Gi	Metabotropic $mGlu_7$	Gi	Serotonin $5HT_{1E}$	Gi
Adrenergic b_1	Gs	Metabotropic $mGlu_8$	Gi	Serotonin $5HT_{1F}$	Gi
Adrenergic b_2	Gs	Muscarinic M_2	Gi	Serotonin $5HT_4$	Gs
Adrenergic b_3	Gs	Muscarinic M_4	Gi	Serotonin $5HT_6$	Gs
Adrenomedulin	Gs	Neuropeptide Y_1	Gi	Serotonin $5HT_7$	Gs
CGRP	Gs	Neuropeptide Y_2	Gi	Somatostatin SST_1	Gi
Cannabinoid CB_1	Gi	Neuropeptide PP_1	Gi	Somatostatin SST_2	Gi
Dopamine D_1	Gs	Opioid m	Gi	Somatostatin SST_3	Gi
Dopamine D_2	Gi	Opioid d	Gi	Somatostatin SST_4	Gi
Dopamine D_3	Gi	Opioid k	Gi	Somatostatin SST_5	Gi
Dopamine D_4	Gi	PACAP	Gs	Vasopressin V_2	Gs
Dopamine D_5	Gs	Prostanoid DP	Gs	VIP_1	Gs
$GABA_B$	Gi	Prostanoid IP	GS	VIP_2	Gs
Galanin	Gi	Prostanoid EP_2	Gs		
Histamine H_2	Gs	Prostanoid EP_3	Gi		

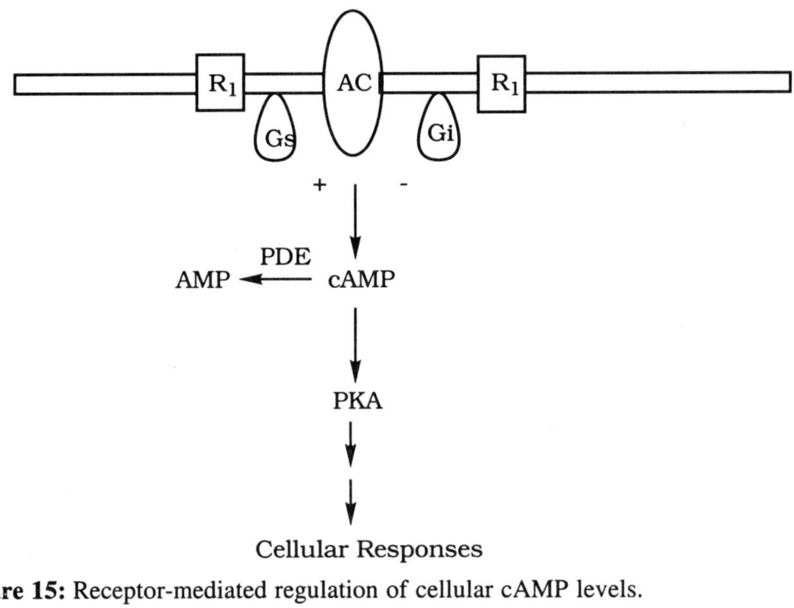

Figure 15: Receptor-mediated regulation of cellular cAMP levels.

2.3.2.2 *High Throughput Functional Assays for Gs- and Gi-Coupled Receptors*

Several commercially available assay methodologies have been developed to measure cAMP levels or adenylyl cyclase activity which do not involve laborious chromatography steps typically required to resolve substrate and product (ATP and cAMP). These new assays are generally amenable to use in high throughput screening. The following section details some of these new assay systems.

2.3.3 Common Cell-Based Receptor Functional Assays

2.3.3.1 *SPA cAMP Assay*

Amersham has developed a commercially available scintillation proximity assay (SPA) kit that can be used for measurement of cAMP. Either whole cells or plasma membranes containing the receptor of interest can be used. The strength of this assay is that laborious separation and washing steps, necessary for traditional measurement methods, are not required resulting in a vast increase in assay throughput. The important components of the SPA cAMP assay system include the use of an anti-cAMP primary antibody and a secondary antibody coupled to SPA beads. The principle of the assay is that trace [^{125}I]cAMP added to the assay reaction mixture will compete with cAMP generated in the assay for binding to the primary antibody. The SPA bead is brought into close proximity of [^{125}I]cAMP through the secondary antibody. As cAMP increases, the signal in the assay decreases.

The steps involved in the Amersham cAMP assay are as follows:

1) Cells or membranes, agonist if desired, and test extracts are added to a standard microtiter plate and incubated for a pre-established period at room temperature.
2) Microtiter plates are then centrifuged and an aliquot is removed and placed in another microtiter plate, along with standards and buffer for blank wells. It may be possible to perform the initial cAMP assay and cAMP determinations in the same plate if low amounts of membranes or cells are used and if the level of cAMP generated in the initial incubation is sufficient to fall within the standard curve. If cells are used, they need to be lysed prior to centrifugation of plates.

3) After transfer of an aliquot of the reaction mixture to a microtiter plate, [^{125}I]cAMP tracer, anti-cAMP antibody, and SPA-anti-rabbit secondary antibody conjugate are added, and plates are incubated for 15-20 hours at room temperature.
4) Plates are then counted in a microtiter plate scintillation counter. cAMP values in experimental samples are determined by extrapolation from the generated cAMP standard curve. Typically, a standard curve is generated from each assay plate.

The components and steps of the SPA-cAMP assay are listed in Figure **16**.

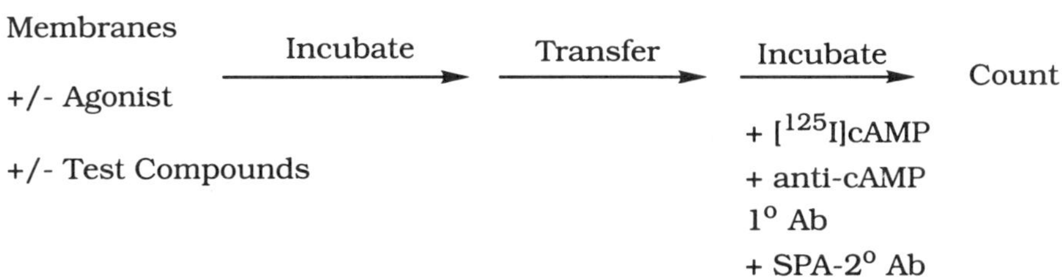

Figure 16: Steps involved in the SPA-cAMP assay.

The following equipment would be required for a high throughput SPA cAMP assay:

-96 well liquid handling system (optional)

-Microtiter plate scintillation counter (Wallac MicroBeta or Packard Top Count)

-robotics system for plate manipulations (optional)

2.3.3.2 FlashPlate cAMP assay

Dupont-NEN has adapted its FlashPlate™ technology to be used as a high throughput assay method for measurement of cAMP levels. The basic principle of the assay is competition of the trace [^{125}I]cAMP added and cAMP generated during incubations for binding to an anti-cAMP antibody which is affixed to scintillant-coated microtiter plate wells. The primary advantage of this assay method, like the SPA assay, is that separation of bound from free tracer radiolabelled cAMP is not required.

The steps involved in the FlashPlate™ cAMP assay are as follows:

1) Cells or membranes, agonists if desired, and test extacts are added to microtiter plates and incubated for a predetermined period at room temperature.
2) Plates are then centrifuged, an aliquot of the reaction mixture is removed and placed in the FlashPlate. It may be possible to perform the initial cAMP generation assay and the cAMP measurement both in FlashPlates if the amount of membrane added is not too expensive and if the levels of cAMP generated fall within the standard cAMP curve. If cells are used, they need to be lysed prior to centrifugation.
3) cAMP reference standards followed by aliquots of the reaction mixture are added to appropriate wells. Tracer cAMP is then added and the plates are incubated for 18-24 hours at 2-8°C.
4) Plates are then read in a microtiter plate scintillation counter. Concentrations of cAMP are extrapolated from the standard curve. Figure **17** details the components and steps associated with the FlashPlate cAMP assay.

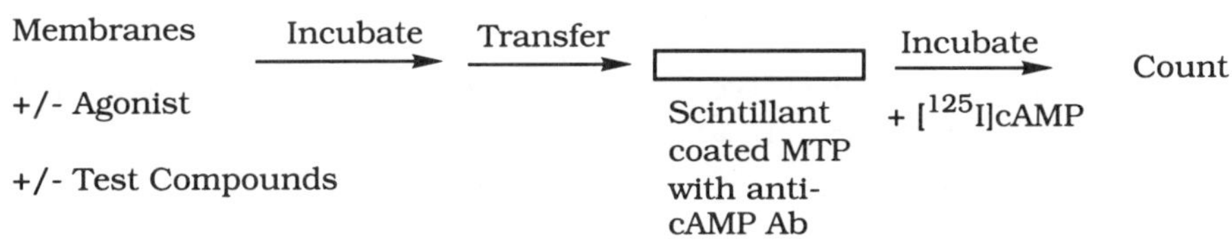

Figure 17: Steps involved in the FlashPlate assay for cAMP.

The following equipment would be required for performing the high troughput FlashPlate cAMP assay:

-96 well liquid handling system (optional)

-Microtiter plate scintillation counter (Packard Top Count only)

-robotics system for plate manipulations (optional)

2.3.3.3 ELISA-Based cAMP Assays

Cayman Chemical has developed a competitive enzyme immunoassay kit for measurement of cAMP levels. The basic principle of this assay is the competition between generated cAMP and a [^{125}I]cAMP tracer (which is linked to an acetylcholinesterase molecule) for a limited number of cAMP-specific rabbit antiserum binding sites. The secondary antibody, a mouse monoclonal anti-rabbit antibody, is previously attached to wells in microtiter plates. The following steps are involved in the Cayman cAMP assay:

1) Cells or membranes, agonist if desired, and test extracts are added to microtiter plates and incubated for predetermined times at room temperature.
2) If cAMP incubations are performed in separate plates, plates are centrifuged and an aliquot of the assay incubation mixture is transferred (and diluted to a level where experimental values fall within the standard curve) into the cAMP determination plate. It may be possible to perform assay incubations in the cAMP determination plate if the concentration of membranes added is not too high and if cAMP generated does not require dilution.
4) cAMP reference standards are added to appropriate wells. Aliquots of the incubation reaction mixture are added to appropriate wells.
3) Aliquots of cAMP-AChE conjugate and primary antibody solutions are then added to plates followed by an 18 hour incubation at room temperature.
4) Plates are then washed five times, Ellman reagent is added, and plates are incubated in the dark for 60-90 minutes before reading the plates at 412 nm.

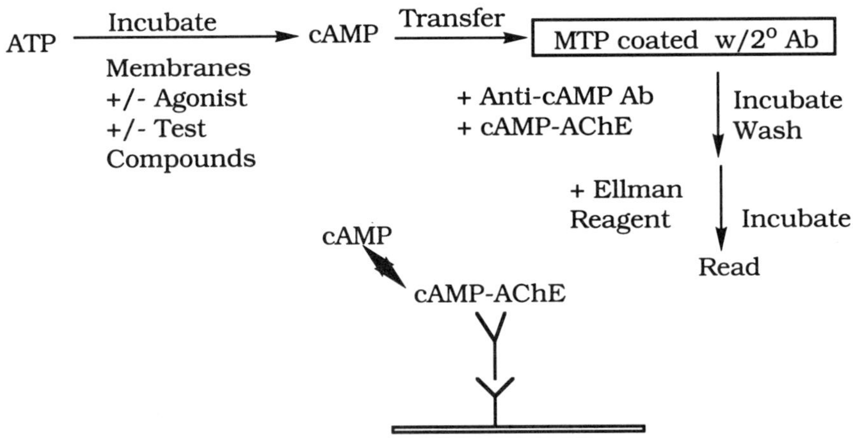

Figure 18: Steps associated with the Cayman ELISA-based cAMP assay.

The product of the enzymatic reaction is a yellow color that can be read in a plate reader at 412 nm. The intensity of the color is proportional to the concentration of cAMP present in the well. The sensitivity of the assay is down to 10 pmol/ml without acetylation. The concentration of cAMP in test samples is determined by extrapolation from a standard curve. The assay steps associated with Cayman RIA assay are illustrated in Figure **18**.

Standard equipment required for performing the Cayman assay are:

-96-well plate liquid handling device (optional)

-microtiter plate reader

-microtiter plate washer

-robotic system for plate manipulations (optional)

Amersham also offers a nonradioactive ELISA-based colorimetric assay kit for measurement of cAMP levels. The primary components of the kit are microtiter plates precoated with a donkey anti-rabbit secondary antibody, a rabbit anti-cAMP antibody, and a tracer cAMP conjugated with peroxidase. cAMP generated in the assay will compete for binding of the cAMP-peroxidase conjugate to the primary antibody. This assay has a sensitivity of 38 pmol/ml using unacetylated standards and 14 pmol/ml using acetylated standards.

The steps involved in the Amersham colorimetric cAMP assay are as follows:

1) Cells or membranes, agonist if desired, and test extracts are added to a microtiter plate followed by incubation for a predetermined period.
2) Microtiter plates are then centrifuged, and an aliquot of the reaction mixture is added to the cAMP determination plate containing secondary antibody. It may be possible to perform the cAMP incubation and determinations in the same plate if small amounts of membranes are used and the amount of cAMP generated falls within the cAMP standard curve.
3) Following transfer of experimental samples and addition of standards and buffer to blank wells, primary anti-cAMP primary antibody is added, and plates are incubated for 2 hours at 3-5 °C.
4) cAMP-peroxidase conjugate is then added to plates and the plates are incubated at 3-5°C for 60 minutes.

5) Plates are then washed, the enzyme substrate is added, and the plates are incubated for an additional 60 minutes at room temperature.
6) The reaction is then stopped by addition of sulfuric acid and the plates are read at 450 nm within 30 minutes. The levels of cAMP in experimental samples are determined by extrapolation from the cAMP standard curve.

Figure **19** details the steps involved in the Amersham EIA cAMP assay.

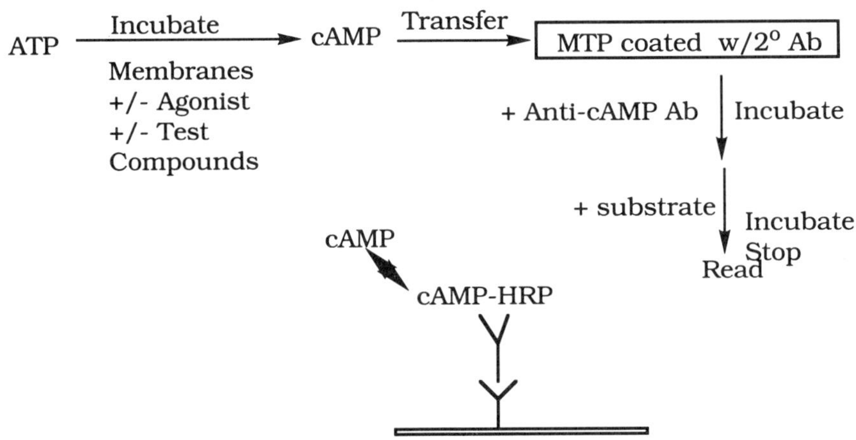

Figure 19: Steps associated with the Amersham ELISA-based cAMP assay.

2.3.3.4 *cAMP Response Element Reporter Gene Assay*

Molecular biological techniques have been used to develop a homogenous cell-based reporter gene assay that is associated with receptor-mediated generation of cAMP (Sassone *et al.*, 1988; Castan'on *et al*, 1994; Migeon *et al.*,, 1994). Specifically, increases in cellular cAMP levels promote the activation of PKA, phosphorylation of the cAMP-response element, enhanced binding of the transcription factor CREB, and enhanced transcription of the reporter gene of choice. Common reporter genes include β-galactosidase and luciferase. Plasmids containing the cAMP response element (CRE) and the luciferase reporter gene have been contructed and transfected into cells containing the Gs or Gi linked receptor of interest. Alternatively, both receptor and CRE-luciferase contructs may be co-transfected into mammalian cells of interest such as CHO, 293,

BHK, and 3T3 cells. A number of pharmaceutical companies have utilized this technology for the development of high throughput screening assays. This assay provides an excellent example of how molecular biology can be used to develop robust and sensitive homogenous functional receptor screening assay. Although preparation of plasmid constructs and optimizing expression levels can be laborious, once developed, these assays serve as excellent primary screens. One previous problem with the luciferase assays concerns short lived luminescence signals. With short-lived signals, it is not possible to initiate and read a whole assay microtiter plate because luminometers read one row at a time. New luciferase substrates that generate a long-lived luminescent signal have been developed which alleviate this problem. Another advance is the development of plate readers that can read all wells simultaneously. Optical cross talk can be a significant problem in luciferase assays and can be reduced by using specialized microtiter plates (Kolb *et al.*, 1996). Figure **20**. details the steps associated with the CRE luciferase reporter assay.

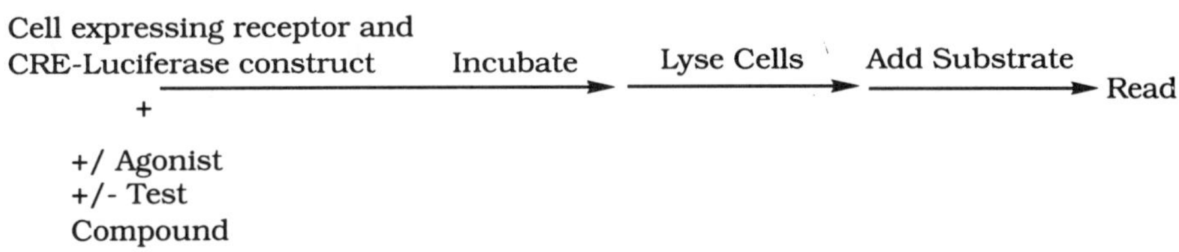

Figure 20: Steps associated with a CRE-luciferase reporter gene assay.

2.3.3.5 Gq-coupled Receptor Signal Transduction Pathways

2.3.3.5.1 Introduction

Many different mammalian G-protein coupled receptors utilize the 1,4,5-inositol triphosphate (IP_3)/diacylglycerol (DAG)/calcium signal transduction pathway. The first step in this signaling pathway involves agonist binding to receptor leading to coupling with a specific heterotrimeric G-protein, Gq (Lee *et al*, 1994) and possibly G_{11} (Nishizuka, 1992). Formation of this high affinity ternary complex then leads to the release of G-protein bound GDP and exchange for GTP, dissociation of G-protein into GTP-bound α and $\beta\gamma$ subunits, and activation of PI-specific phospholipase C (PLC). Experimental evidence suggests that the PLC β isozymes are activated by G-protein subunits (Taylor et al., 1991; Lee et al., 1992; Jhon et al., 1993; Lee et al., 1994).

Furthermore, it appears that Gq α-subunit activates PLCβ1and PLC β3 while the Gi βγ stimulates the PLCβ2 isoform. Gq subunit activation of PLCβ is terminated by the hydrolysis of GTP bound to the α–subunit to GDP, which in turn leads to reassociation of G-protein subunits.

G-protein-mediated stimulation of PLCβ results in hydrolysis of membrane associated phosphatidylinositol species (phosphatidylinositol, phosphatidyl-4-phosphate, and phosophatidyl-4,5-bisphosphate) resulting in the formation of inositol phosphate species (inositol-1-phosphate, inositol-1,4-diphosphate, and inositol-1,4,5-triphosphate) and diacylglycerol (DAG) (Berridge, 1993). Inositol-1,4,5-triphosphate (IP_3) formed then binds to a specific IP_3 receptor located on the endoplasmic reticulum membrane resulting in the release of intracellular stores of calcium. This receptor-mediated increase in intracellular calcium is transient and returns to basal levels in about 10-30 seconds. In many cells, the depletion of intracellular calcium stores in turn activates a capacitance calcium channel (also known as CRAC) which allows the influx of extracellular calcium resulting in a more sustained increase in intracellular calcium (Putney, 1990; Hoth *et al.*, 1992). Activation of voltage-dependent L-type calcium channels in certain cells can also lead to influx of

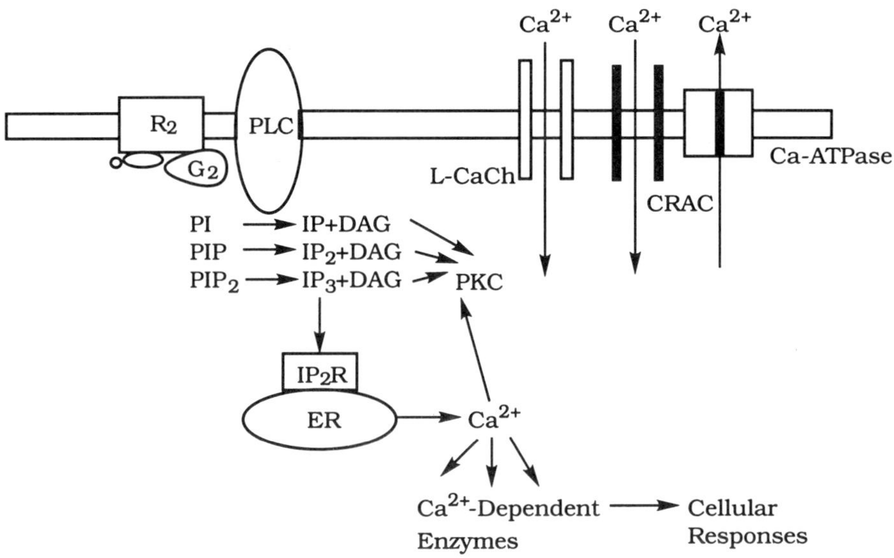

Figure 21: Receptor-mediated regulation of intracellular calcium.

extracellular calcium and contribute to increases in intracellular calcium. The combination of increases in intracellular DAG and calcium leads to activation of protein kinase C (PKC) and phosphorylation of specific cellular substrates as well as activation of a number of calcium-dependent enzymes. The levels of intracellular calcium are restored to basal levels by the combined actions of calcium-ATPase and the Na^+/Ca^{2+}antiporter. Figure **21** provides a diagrammatic representation of cellular components involved in receptor-mediated stimulation of PLC and increases in intracellular calcium.

There are a large number of G-protein coupled receptors which activate PLC and promote increases in intracellular calcium. Table **4** lists some representative Gq-coupled receptors.

Table-4: Gq Coupled Receptors that Mediate Increase in Intracellular Calcium

Adrenergic α_{1A}	Chemokine CC CK_1	Interleukin 8 IL-8_B
Adrenergic α_{1B}		Leukotriene B_4
Adrenergic α_{1D}	Chemokine CC CK_2	Leukotriene C_4
Angiotensin AT_1	Chemokine CC CK_3	Leukotriene D_4
Bombesin BB_1	Cholesystekinin CCK_A	Melatonin ML_2
Bombesin BB_2	Cholesystekinin CCK_B	Metabotropic $mGlu_1$
Bombesin BB_3	Endothelin ET_A	Metabotropic $mGlu_5$
Bradykinin B_1	Endothelin ET_B	Muscarinic M_1
Bradykinin B_2	Histamine H_1	Muscarinic M_3
	Interleukin 8 IL-8_A	Muscarinic M_5
MCPI	Prostanoid EP_3	Tachykinin NK_2
Neurotensin	Purinoreceptor P_2	Tachykinin NK_4
Oxytocin OT	Purinoreceptor P_{2U}	Thrombin
PAF	Serotonin $5HT_{2A}$	Vassopressin V_{1A}
Prostanoid FP	Serotonin $5HT_{2B}$	Vassopressin V_{1B}
Prostanoid TP	Serotonin $5HT_{2C}$	
Prostanoid EP_1	Tachykinin NK_1	

In addition to Gq coupled receptors, some receptors that couple to Gi have been reported to alter levels of intracellular calcium. These receptors, through a G-protein dependent mechanism, reduce or inhibit calcium currents in certain cells. A list of these receptors is shown in Table **5**.

Table 5. Non Gq G-Protein Coupled Receptors That Mediate Reduction in Intracellular Calcium

Adenosine A_1
Adrenergic a_{2A}
Adrenergic a_{2B}
Cannabinoid CB_1
Cannabinoid CB_2
Dopamine D_2
Dopamine D_3
$GABA_B$
Galanin
Neuropeptide Y_2
Opioid m
Opioid d
Opioid k

2.3.3.5.2 High Throughput Assays For Gq-coupled Receptors

Measurement of agonist-mediated increases in either intracellular IP3, DAG, or calcium concentrations can serve as a functional assay for Gq coupled receptors. Although assay kits are available for measurement of IP3 and DAG, these assays are not amenable to high throughput screening because they involve laborious extractions of cells with organic solvents.

Assay of receptor-mediated changes in intracellular calcium is more suitable for development of a high throughput functional receptor screening assay. There are two approaches that can be taken to measure changes in intracellular calcium. One method consists of measuring receptor-mediated influx of [$^{45}Ca^{2+}$] into appropriate cells containing the receptor of interest. Although this assay method has been used for receptor studies, it is not suitable for high throughput screening for several reasons. First, receptor activation must result in the significant influx of calcium ions in addition to release of intracellular stores. Second, the signal-to-noise ratio associated with radioactive ion flux assays is generally poor. Third, this method requires adherent cells because unincorporated [$^{45}Ca^{2+}$] needs to be removed by washing cells. Due to these considerations, this assay method is not generally suitable for high throughput screening.

Another method for measuring changes in intracellular calcium involves the use of calcium sensitive fluorescent dyes. A number of different dyes have been developed which are sensitive to low nanomolar changes in intracellular calcium and include Fura-2 AM, Fluor-3 AM, Indo 1 AM, and Calcium Green AM. One important feature of these dyes is that once they enter cells during loading, cellular esterase activity cleaves an ester moiety contained within the dye resulting in intracllular trapping of the dye. These calcium sensitive dyes have been widely used for evaluation of receptor-mediated changes in intracellular calcium for a number of years. The technique involves the use of a standard fluorometer and experiments are performed using a single quartz cuvette. Studies with numerous receptor systems have indicated that receptor-mediated increases in intracellular calcium are transient and generally last for 10 to 30 seconds before returning to basal levels. Because of this transient nature, the assay could not be configured into a microtiter plate format. Standard fluorescence plate readers are capable of reading only one row of the plate at a time and the calcium signal can not be captured for the whole microtiter plate because of this transient response. A recent new advance in fluorescence plate readers has now made it practical to perform intracellular calcium assays in a microtiter plate format. This advance consists of development of real time fluorescence plate readers that also contain liquid handling devises (Schroeder *et al.*, 1996). With these new readers, fluorescence reading is started prior to addition of receptor agonists. This, combined with the ability to read all wells at once, has allowed the capture of these transient calcium signals initiated by agonist binding to receptors. The steps associated with an intracellular calcium assay using fluorescent calcium sensitive dyes and a real time plate reader are shown in Figure **22**.

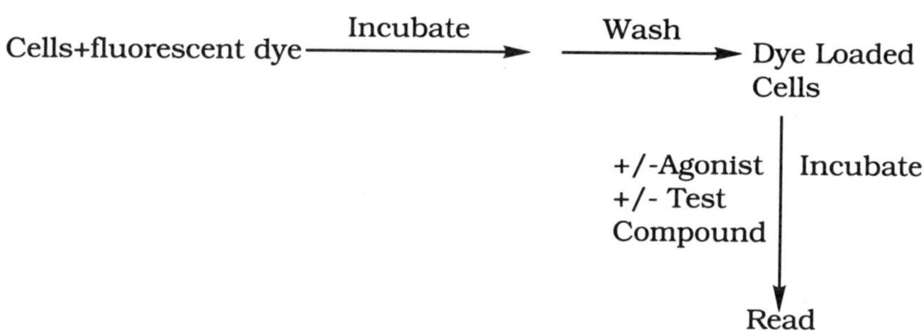

Figure 22: Steps involved in a fluorescence-based intracellular calcium assay.

The intracellular calcium assay can also serve as an assay to identify compounds that interact with signal transduction molecules associated with receptor-mediated increases in intracellular calcium. These signal transduction molecule targets could include Gq, PLCβ, the IP3 receptor, and possibly L-type calcium channels and the CRAC channel. Since multiple targets are associated with the assay, active extracts need to be evaluated to determine the locus of action in the signal transduction pathway leading to increases in intracellular calcium.

2.3.3.6 High Throughput Functional Assays for Ligand-gated Ion Channels

Agonist occupation of a number of ligand-gated ion channels can lead to increases in intracellular calcium. This can occur due to either calcium ion permeability of the ligand-gated channel or to depolarization of the plasma membrane leading to opening of L-type calcium channels. For example, NMDA excitatory amino acid receptors are permeable to both Na^{+} and Ca^{+2} ions and increases in intracellular calcium with NMDA activation have been well documented (Ascher *et al.*, 1988). Increases in intracellular calcium accompanying NMDA binding are due both to calcium entry through the NMDA channel as well as opening of L-type calcium channels due to membrane depolarization resulting from sodium ion influx. Related AMPA and kainate receptors are not permeable to calcium ions but do allow influx of sodium ions into cells leading to membrane depolarization and opening of L-type calcium channels. Table **6** lists ligand-gated ion channels for which increases in intracellular calcium have been reported. High throughput functional assays for these receptors can be developed using calcium sensitive dyes and a real time fluorescence plate reader as described for Gq-coupled receptors.

Table 6. Ligand Gated Ion Channels Reported to Alter Intracellular Calcium Concentrations

Receptor	**Receptor**
AMPA	Nicotinic ACh (α7)
Kainate	Purinoreceptor P_2X1
NMDA	Purinoreceptor P_2X2
Nicotinic ACh (muscle)	Purinoreceptor P_2X3
Nicotinic ACh (ganglionic)	Purinoreceptor P_2X4
Nicotinic ACh (neuronal CNS)	Purinoreceptor P_2Z

2.3.3.7 Receptor Protein Tyrosine Kinase Functional Assays

2.3.3.7.1 Introduction

Many cellular activities of mammalian cells are controlled by receptor protein tyrosine kinases (RPTK's). These tyrosine kinases are involved in the regulation of a number of cellular programs; the most notable are the control of cell growth and differentiation. Because of these properties, tyrosine kinases are thought to be important in the genesis of many neoplasias. RPTK's are unique in that they not only possess a ligand binding site but also contain intrinsic tyrosine kinase enzymatic activity. Examples of RPTK's include epidermal growth factor (EGF), platelet-derived growth factor (PDGF), fibroblast growth factor (FGF), and insulin receptors.

2.3.3.7.2 RPTK-Associated Signal Transduction Pathways

RPTK's activate complex signal transduction pathways that regulate the activity of nuclear transcription factors and ultimately regulate gene transcription. The signal transduction pathways activated by RPTK's generally include the Ras pathway and the MAP kinase cascade.

The first step in the signal transduction pathway subsequent to ligand binding involves a ligand-mediated dimerization of the receptor (Schlessinger, 1988; Williams, 1992; Ueno *et al.*, 1991; Overduin et al., 1992). Receptor dimerization is then accompanied by autophosphorylation of intracellular receptor subunits leading to activation of the cytoplasmic tyrosine kinase catalytic domain (Schlessinger, 1988). The next step is the specific interaction of the autophosphorylated receptor with specific signal transduction molecules.

Protein-protein interactions between autophos-phorylated RPTK's and signal transduction molecules are governed by a very specific interaction of RPTK phosphotyrosine residues with SH2 (src homology) domains contained within the sequence of signal transduction molecule (Koch *et al.*, 1991; Heldin, 1991; Margolis, 1992). SH2 domains comprise approximately 100 amino acid residues that recognize very restricted sequences surrounding phosphotyrosine residues on RPTK's. Based on both structural and biochemical studies, it appears that 3-5 amino acid residues C-terminal to the phosphotyrosine interact with the surface of the SH2 domain. Another protein-protein interaction domain found in several substrates for RPTK's, is the SH3 domain (Pawson, 1992; Musacchio *et al.*, 1992; Mayer *et al*, 1993). SH3 domains consist of approximately 60 amino acids and bind to ten amino acid proline-rich binding sites contained within specific proteins.

Some signal transduction molecules that have been found to directly associate with activated mammalian RPTK's through SH2 and/or SH3 domains include PLCγ, PLD, and PI3-kinase. Several activated RPTK's have been known to stimulate membrane PI turnover leading to increases in IP3, DAG, and calcium. This effect can be explained by the observation that activated RPTK's interact with PLCγ via a SH2 interaction, phosphorylate and activate PLCγ, leading to PI turnover and generation of IP3, DAG, and increases in intracellular calcium. Another dominant SH2-mediated interaction with activated RPTK's is the interaction with PI3-kinase (Escobedo *et al.*, 1991). Recent evidence suggests that the 85 kDa subunit of PI3-kinase may be phosphorylated resulting in reduced interaction with RPTK's and may serve as a negative feedback mechanism for activation of PI3-kinase (Carpenter *et al.*, 1993).

RPTK's also associate with signal transduction molecules that do not have an enzymatic activity but act as adapters. The interaction of these adapter proteins with RPTK's is also governed by SH2 domain-phosphotyrosine and SH3 recognition motifs. Some examples of mammalian adapter proteins involved in RPTK signal transduction include CRK, NCK, Grb2, and SHC (Van der Geer *et al*, 1994; Van der Geer *et al.*, 1995).

The small GTP-binding protein, Ras, also plays an important role in signal transduction pathways activated by RPTK's. Ras normally cycles between an inactive GDP-bound form and an active GTP-bound form. Because the dissociation of GDP from Ras is very slow, Ras tends to accumulate in an inactive GDP-bound state. To activate Ras, an auxiliary protein, termed a guanine nucleotide exchange factor (GNRF), is required to facilitate the exchange of GDP for GTP. The specific mammalian protein, hSos, is a GNRF that plays an important role in activation of Ras. Activated Ras is inactivated by the hydrolysis of bound GTP to GDP which is commonly referred to as the GTPase step. The GTPase reaction is catalyzed by a GTPase activating protein called GAP (Gibbs *et al.*, 1988). GAP interacts with a number of RPTK's via its SH2 domains and is phosphorylated on tyrosine residues.

Recently, it has been shown that Ras-GTP directly interacts with the ser/thr protein kinase, Raf-1. In addition to interaction with Raf-1, it also appears that phosphorylation of Raf-1 is also necessary for Raf-1 activation and that this occurs once Raf-1 is translocated to the plasma membrane as a result of binding to Ras-GTP. Thus, the role of activated Ras in RPTK signal transduction is to recruit the ser/thr kinase, raf-1, to the plasma membrane, where a tyrosine kinase-generated signal fully activates the membrane bound raf-1 (Williams *et al.*, 1992). Raf-1 in turn, activates a protein kinase cascade highly conserved throughout evolution. According to our current knowledge, MAP kinase kinase (also known as MEK), a dual specificity kinase, is activated by either MAP kinase kinase kinase (also known as MEKK), Raf-1, or MOS by ser/thr phosphorylation. Activation of MEK by raf-1 involves phosphorylation of two ser residues that are four amino acids apart in kinase subdomain VIII (Alessi *et al.*, 1994). MEK then promotes the thr and tyr phosphorylation of MAP kinase leading to its activation. It has also been suggested that PKC can phosphorylate and activate both Raf-1 and MEKK (Kolch *et al.*, 1993). Once MAP kinase is activated, it can translocate into the nucleus, where it can phosphorylate and activate transcription factors such as Elk-1 (Treisman, 1994). Other transcription factors activated by MAP kinase include c-Jun and c-Fos. Therefore, MAP kinase serves as a relay between cytoplasmic and nuclear events. In addition to transcription factors, another set of substrates for MAP kinase are upstream proteins of the MAP kinase cascade. Examples are the NGF receptor, the EGF receptor, PTP2C, Sos, Raf-1, and MEK. Therefore, MAP kinases may also play a role in a feedback mechanism for upstream components. MAP kinases also phosphorylate cytoskeletal proteins such as MAP-1, MAP-2, MAP-4, and Tau.

2.3.3.7.3 High Throughput Functional Assays For RPTK's

There are several different assays that could serve as a functional screening assay for RPTK's. These could include measurements of phosphorylation of specific signal transduction molecules by the RPTK as described above, activation of Ras, or MAP kinases, or activation of specific transcription factors. Of these assays, only transcription assays are amenable to high throughput screening. Development of a transcription assay would involve construction of a plasmid containing a suitable transcription factor response element and a reporter gene such luciferase or β-galactosidase. Suitable cells would then have to be transfected with this gene construct. Aside from transcription assays, there are only a couple of other suitable choices for a high throughput functional RPTK assay and these are detailed below.

2.3.3.7.4 Intracellular Calcium Assay

As stated above, several RPTK's stimulate PLCγ resulting in elevations of IP3, DAG, and intracellular calcium. The same high throughput intracellular calcium assays described for Gq coupled receptors can also be adapted as functional assays for RPTK's.

2.3.3.7.5 Proliferation Assays

RPTK's promote proliferation of a number specific human cells. Therefore, simple proliferation assays can serve as functional receptor assays for RPTK's. The first task in developing such an assay is finding the appropriate human cell that contains the RPTK of interest. If this can be accomplished, a number of different methods are available to measure cellular proliferation (DNA synthesis or metabolic activity) including radiometric assays in which uptake of [^{3}H]thymidine is measured, colorimetric assays using BrdU, MTT, XTT, or WST-1, or fluorometric assays (Alamar-Blue) measuring incorporation of fluorescent tagged purines or pyrimidines (Heeg *et al.,* 1985; Denizot *et al.,* 1986; Gerlier *et al.*, 1991; Perros *et al.*, 1991; Weichert *et al.*, 1991; Perros *et al.*, 1991; Weichert *et al.*, 1991; Bank *et al.*, 1991; Tajima *et al.*, 1992; Buttke *et al.*, 1993; Van Weerden *et al.*, 1993; Gressner *et al.*, 1996; Wagner *et al.*, 1997; Mgbonyebli *et al.*, 1998). These assays can be configured in a high throughput microtiter plate format. One potential caveat of using cellular proliferation assays for screening natural product extracts is that they generally require lengthy incubations in which false positive actives would be generated due to possible nonspecific cytotoxicity of extracts. In addition, determining the locus of action of an active extract in this assay could be complex due to the large number of signal transduction molecules that are associated in the signal transduction pathway associated with receptor-mediated proliferation. To serve as examples of proliferation assays, several proliferation assay kits that are commercially available are from Calbiochem detailed below.

a) [^{3}H]Thymidine Incorporation

A popular proliferation assay based on DNA synthesis is measurement of [^{3}H]thymidine incorporation into DNA. During cell proliferation, DNA has to be replicated before the cell divides into two daughter cells. The close association between DNA synthesis and cell doubling is the basis of the of [^{3}H]thymidine incorporation proliferation assay. The major disadvantage of using the [^{3}H]Thymidine incorporation assay is associated with handling and disposal of radioactive waste generated by the assay. The steps associated with this type of assay are listed in Figure **23**.

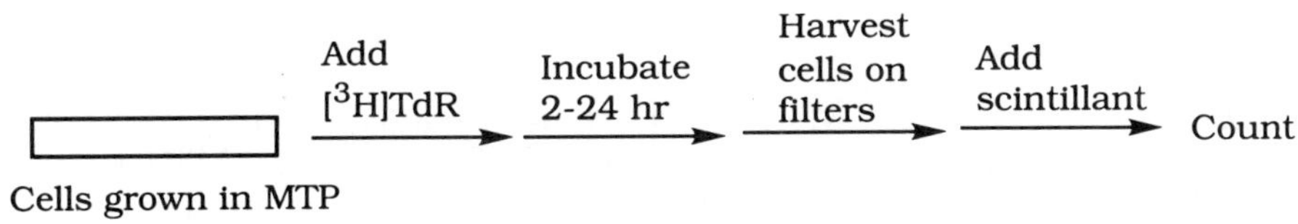

Figure 23: Steps associated with the [^{3}H]thymidine incorporation proliferation assay.

b) BrdU Assay

The thymidine analog, 5-bromo-2-deoxyuridine (BrdU), has been used by Calbiochem to develop a nonradioactive proliferation assay kit. The basis of the assay is that proliferating cells will have increased levels of DNA synthesis and incorporate BrdU into DNA. BrdU incorporation into DNA is measured using a monoclonal antibody that is conjugated with peroxidase. Either a colored peroxidase substrate (TMB) or a chemiluminescent substrate (luminol/4-iodophenol) can be used as a substrate for peroxidase. Advantages of the BrdU assay are that the assay can be performed in one microtiter plate and that it is nonradioactive. One of the possible disadvantages of the BrdU assay using TMB as a peroxidase substrate is that the assay may not be linear over a broad range of cell proliferation due to the use of a plate reader. This is not a problem when using the chemiluminescent substrate and a luminometer. The assay steps associated with this assay are shown in Figure **24**.

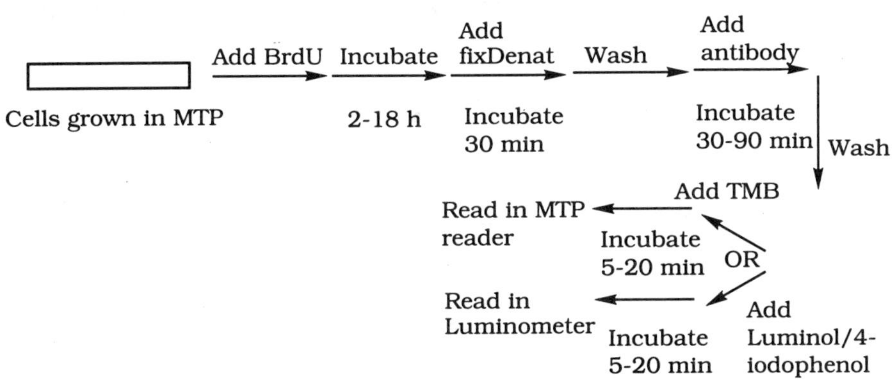

Figure 24: Steps associated with the BrdU proliferation assay.

c) MTT Assay

Calbiochem also offers a MTT proliferation assay kit which measures cellular metabolic activity. The basis of the MTT assay is that MTT is only cleaved to form colored formazan dye by living cells with a viable succinate-tetrazolium reductase system. After solubilization of cells, the colored formazan product can be easily measured by reading plates at 570 nm in a plate reader. The advantages of the MTT assay are that the assay can be performed in one microtiter plate and that it is nonradioactive. The steps associated with the MTT assay are shown Figure **25**.

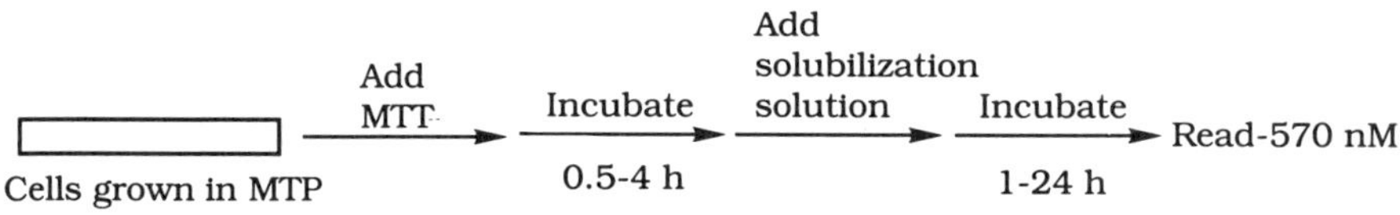

Figure 25: Steps associated with the MTT proliferation assay.

d) XTT Assay

The XTT proliferation assay kit offered by Calbiochem measures cellular metabolic activity similar to the MTT assay. The assay is based on the cleavage of the tetrazolium salt of XTT by living cells in the presence of an electron coupling reagent. The reaction product is a formazan salt which is quantitated by reading plates at 570 nm in a plate reader. The advantages of the XTT assay are that it can be performed in one plate, the reaction product is soluble, multiple time points can be made in one assay, and it is a nonradioactive assay. The steps associated with the assay are shown in Figure **26**.

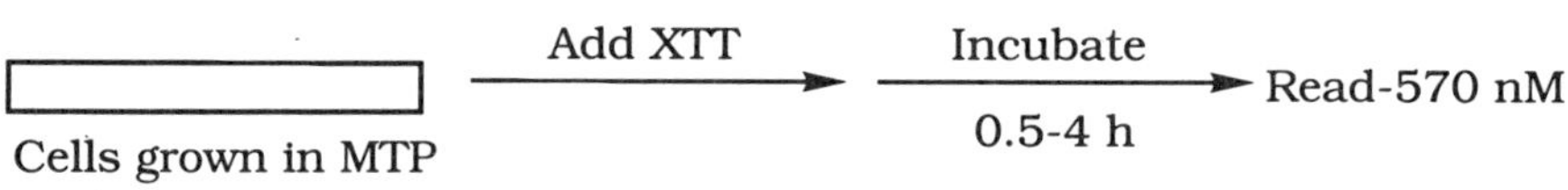

Figure 26: Steps associated with the XTT proliferation assay.

e) WST-1 Assay

The WST-1 proliferation assay kit offered by Calbiochem is similar to the XTT assay in that it measures cellular metabolic activity and the product generated is a formazan salt which is measured by reading plates at 570 nm with a plate reader. The advantages of the WST-1 assay is that the assay can be performed in one plate, the reaction product is soluble, repeated measurements can be made in the assay, and it is nonradioactive. The assay steps associated with the WST-1 assay are shown in Figure **27**.

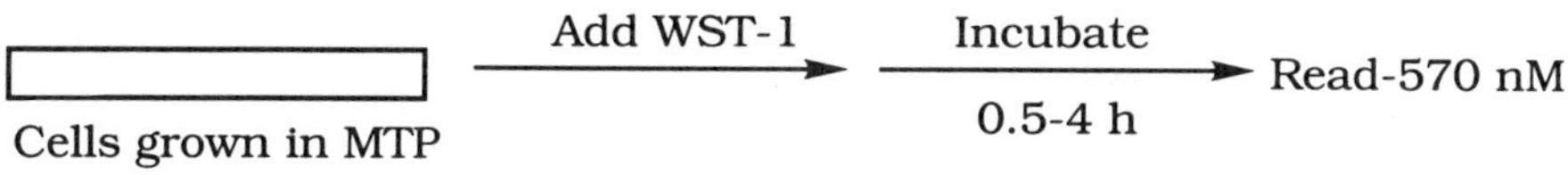

Figure 27: Steps associated with the WST-1 proliferation assay.

2.3.3.8 Cytokine Receptor Functional Assays

2.3.3.8.1 Cytokine Receptor Signal Transduction Pathways

Recent studies suggest that the cytokine receptor family may share several features with the tyrosine kinase receptor family including tyrosine kinase activation, mitogenesis, activation of the Ras-MAP kinase pathway, and activation of PI-3-kinase. Agonist binding to the majority of cytokine receptors has also been shown to result in activation of a kinase cascade involving Jak kinases which ultimately results in phosphorylation and activation of specific cellular transcription factors called STAT proteins. The STAT family of transcription factors regulate the activation or inactivation of immediate early genes in cytokine responsive cells.

2.3.3.8.2 Functional Assays for Cytokine Receptors

Several assays could be selected to serve as a functional assay for cytokine receptors. These could include assays for activation of JAK kinases, PI3-kinase, Ras, or MAP kinases. However, these particular assays would not be suitable for high throughput screening because they would be labor intensive. There are primarily two types of high throughput functional assays that could be envisioned for cytokine receptors. The simplest cytokine receptor assay would consist of measuring cytokine induced increases in proliferation of appropriate cells which contain the receptor of interest. Cellular proliferation could be quantitated in a 96-well microtiter plate format by the same methodologies as discussed for RPTK's. Alternatively, reporter gene assays could be developed using STAT response elements coupled to reporter genes such as luciferase or β-galactosidase. Since both proliferation and transcription assays involve receptor-mediated activation signal transduction pathways, the locus of action of an active extract could be associated with any of the signal transduction molecules associated with these pathways. It is also important to point out that both proliferation and transcription assays generally require lengthy incubations (8-24 hours) in which there would be a high potential for false positive active extracts due to nonspecific cellular cytotoxicity.

2.3.3.9 Mixed Receptor Functional Assay

2.3.3.9.1 Melanophore Functional Receptor Assay Technology

A novel high throughput functional receptor assay system using frog skin melanocytes has been developed by Michael Lerner at Yale which can be widely applied to all G-protein coupled receptors and RPTKs (Lerner, 1994). The "Melanophore Assay Technology" exploits the natural response of pigment-bearing frog melanocytes which undergo rapid optical density changes mediated by G-protein coupled and receptor protein tyrosine kinases. This receptor-mediated pigment redistribution can be captured

Table 7. List of Receptors in which the Melanosome System has been Used as a Functional Assay System

Gs Coupled	*Gi Coupled*	*Gq Coupled*	*RPTK's*
β_2-Adrenergic	α_2-Adrenergic	α_1-Adrenergic	Epidermal Growth Factor
β_1-Adrenergic	Cannabinoid CB_1	Angiotensin II	
Dopamine D_1	Cannabinoid CB_2	Bradykinin	Platelet-Derived Growth Factor
Dopamine D_1	Dopamine D_2	Endothelin ETA	
Prostaglandin EP_2	Dopamine D_3	Endothelin ETB	
Prostaglandin I_2	Dopamine D_4	Endothelin ETC	
	Muscarinic M_4	Platelet Activating Factor	
	Neuropeptide Y	Tachykinin NK_1	
	δ-opioid	Tachykinin NK_2	
	κ-opioid	Thrombin	
	μ-opioid	Thromboxane	
	MOP2 opioid		
	Prostaglandin EP_3		
	Somatostatin $SRIF_1$		
	Somatostatin $SRIF_{2A}$		
	Somatostatin SST_3		
	Somatostatin SST_4		
	Somatostatin SST_5		

using a standard plate reader. Mammalian receptors transfected into melanocytes "hijack" the natural pigment control system of the cells. Ligands that activate transfected receptors cause rapid aggregation or dispersion in cells depending on the signal transduction system employed by the receptor. The melanophore assay system has been used to develop functional homogenous receptor assays for over 37 different human receptors (see Table **7**). No engineering of receptors or cells is required and both transient and stable transfected receptors can be utilized. Advantages of the melanophore assay are that it is applicable for a number of different receptors, the assay has a rapid readout time (30-60 min), readings are directly with living cells, readout utilizes the natural system and does not require transcription of reporter genes, and it is a nonradioactive assay. This assay technology can be licensed for use in drug discovery programs from Bunsen Rush, Inc.

a) Gs- and Gi-Coupled Receptor Regulation of Melanosome Pigment Distribution

Agonist occupation of endogenous or expressed Gs coupled receptors in melanocytes pretreated with melatonin (to promote melanosome aggregation), leads to rapid pigment dispersion (Abe *et al*, 1969a,b; Potenza *et al*, 1991; Potenza *et al*, 1994). The exact mechanism by which

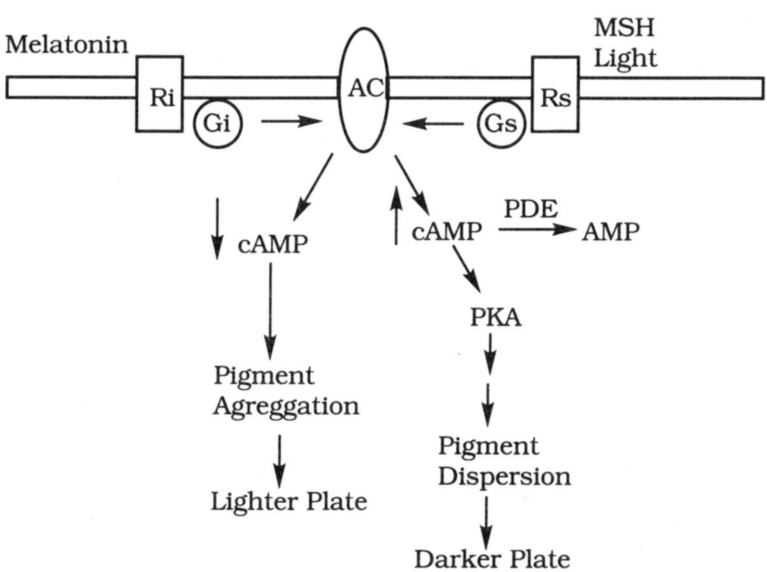

Figure 28: cAMP regulation of melanophore pigment dispersion.

PKA activation leads to melanophore dispersion is currently unclear. One important step is PKA-mediated phosphorylation of a 57-kDa protein associated with pigment organelles. Agonist occupation of endogenous or expressed Gi coupled receptors on frog melanocytes pretreated with light or MSH (to stimulate pigment dispersion) will promote pigment aggregation (McClintock *et al.*, 1993; Potenza *et al.*, 1993; Ebisawa *et al.*, 1994). This response appears to relate to a reduction in the phosphorylated state of the above mentioned 57-kDa protein associated with pigment organelles.

Figure **28** details the signal transduction pathways associated with Gs and Gi -coupled receptor-mediated changes in melanophore pigment dispersion.

b) Gq-Coupled Receptor-Mediated Control of Melanosome Pigment Dispersion

Agonist occupation of endogenous or expressed Gq-coupled receptors in frog melanocytes leads to dispersion of pigment in melatonin treated cells (Sugden *et al.*, 1992). This response has been clearly shown to be dependent upon generation of DAG and activation of PKC. However, the

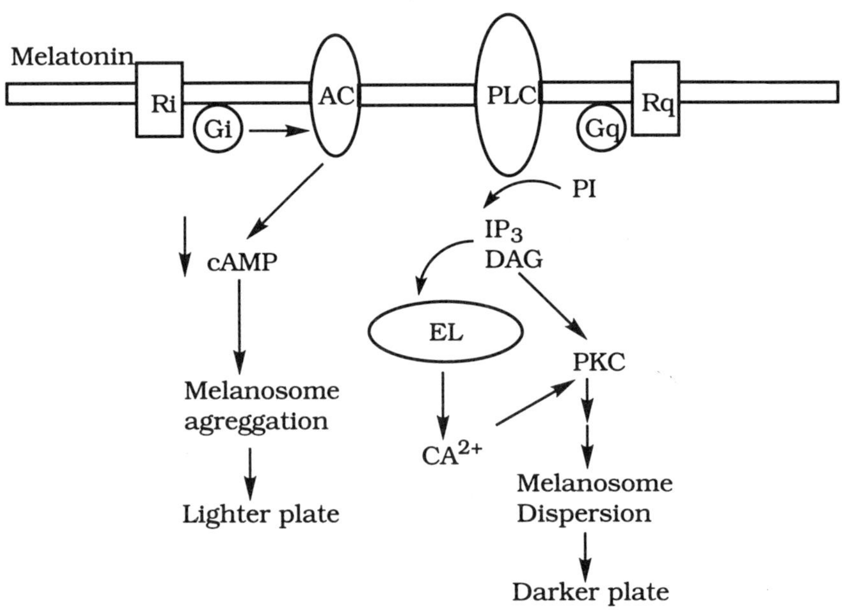

Figure 29: Gq-coupled receptor-mediated melansome dispersion.

mechanism of how PKC activation mediates pigment dispersion is currently unclear. The rapid movement of pigment has been shown to involve the phosphorylation of a 53-57 kDa protein (Rozdzial *et al.*, 1986) and redistibution of pigment granules can be accomplished by applying the protein phosphatase inhibitor, okadaic acid.
Figure **29** details the steps associated with Gq-coupled receptor-mediated stimulation of melanosome dispersion.

c) RPTK Regulation of Melanosome Dispersion

The melanophore system is also applicable as a functional assay for RPTK's which activate PLCγ and ultimately PKC by increasing cellular levels of DAG and intracellular calcium. Growth factor binding to RPTK's that have been transfected into frog melanocytes pretreated with melatonin leads to pigment dispersion. The RPTK-mediated pigment dispersion can be blocked by specific inhibitors of PKC. The steps subsequent to activation of PKC leading to pigment dispersion are currently unknown. Figure **30** details the steps that are associated with RPTK-mediated changes in melanocyte distribution.

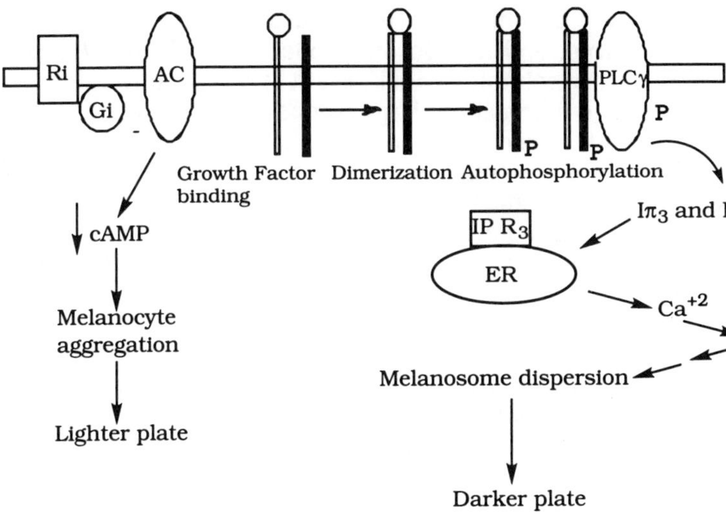

Figure 30: RPTK-mediated melanosome dispersion.

2.3.3.10 Receptor Selection and Amplification Functional Receptor Assay Technology

The receptor selection and amplification technology (RSAT), which has been developed by Mark Brann at the University of Vermont, can be used to develop homogenous receptor functional assays for a wide variety of mammalian receptors including G-protein coupled receptors, RPTKs, and cytokine receptors. A new company, called Receptor Technologies Inc., has been formed to commercialize this technology. The RSAT technology is based on the experimental observation that certain mammalian receptors are able to behave as ligand-dependent oncogenes (Julius *et al*, 1989; and Allen *et al.*, 1991). In developing receptor functional assays, mammalian cells such as NIH 3T3 cells are transiently cotransfected with receptor cDNA of interest and the β-galactosidase gene using the calcium phosphate precipitation method. Agonists binding to expressed receptors promote a selective growth advantage which is measured by an increase in β-galactosidase activity. Non-transfected cells are contact inhibited in the absence of agonist (Messier *et al.*, 1995; Tecle *et al.*, 1995; Brauner-Osborne *et al.*, 1995; Gil *et al.*, 1995).

The RSAT assay system is applicable to a wide range of receptors that utilize a diversity of signal transduction mechanisms including G-protein coupled receptors, tyrosine kinase receptors, JAK/STAT linked cytokine receptors, and transforming oncogenes (Table **7**). The primary advantages of the technology include: it is a high throughput functional assay amenable to automation, large batches of cells can be transfected and then frozen and stored when needed, and multiple receptors can be transfected into cells to create an agonist screen for several receptors at one time. One potential concern with the RSAT assay is that incubations take place over a 4-5 day period. The potential for cytotoxicity of natural product extracts in the assay over this extended incubation period could be considerable.

Receptor Technologies now sells several RSAT kits for different human receptors that can be used for screening natural product extracts. Assay kits are available for the majority of the receptors listed in Table **8**. The assay steps involved in the RSAT assay are given in Figure **31**.

Table 8. Applications of the RSAT Assay System

Gs Coupled	Gs Coupled	Gq Coupled	RPTC's	JAK/STAT
Dopamine D_3	α_{2A}-Adrenergic	α_{1A}-Adrenergic	Trk A,B,C	GM-CSF
	α_{2B}-Adrenergic	α_{1B}-Adrenergic	PDFG	
	α_{2C}-Adrenergic	α_{1C}-Adrenergic		
	Dopamine D_2	Angiotensin AT1		
	Dopamine D_2	Bombesin BB1		
	Muscarinic M_2	Bombesin BB2		
	Muscarinic M_4	Histamine H1		
	μ-Opioid	Interleukin-8A		
	κ-Opioid	Endothelin ETA		
	Prostanoid Ep_3	Endothelin ETB		
	Serotonin $5HT_{1A}$	Muscarinic M_1		
	Serotonin $5HT_{1D}$	Muscarinic M_3		
		Neurokinin NK1		
		Neurokinin NK2		
		Neurokinin NK3		
		Prostanoid EP_1		
		Prostanoid FP		
		Prostanoid TP		
		Serotonin $5HT_{2A}$		
		Vasopressin V_{1A}		
		Vasopressin V_{1B}		

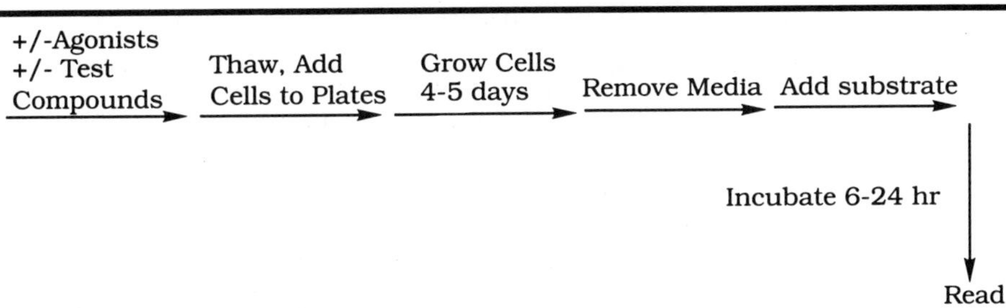

Figure 31: Steps associated with the RSAT assay.

2.4 RADIOLIGAND BINDING ASSAYS

2.4.1 Introduction

Human receptors have been shown to be clearly associated with a multitude of human diseases and disorders. Thus, these receptors have become molecular targets for drug discovery. There are several different structurally related superfamilies of human receptors including G-protein coupled receptors, cytokine receptors, receptor protein tyrosine kinases, ligand-gated ion channels, and steroid receptors and members from all of these superfamilies have been targeted in drug discovery programs. Screening for agonists and/or antagonists of various receptors has traditionally been accomplished by using radioligand binding assays. These assays are relatively simple to perform and are based upon the principles of a competitive binding assay in which the radiolabeled ligand competes with the unlabeled ligand for binding to a receptor in a cell or on the cell surface membrane. Thus, compounds or natural product extracts are evaluated for their ability to compete with a known ligand and this competition is measured by a decrease in the binding of the labeled known ligand. The focus of this section is to describe important aspects of developing and validating receptor radioligand binding assays used to screen combinatorial chemistry reaction products and natural product extracts.

2.4.2 Important Considerations for Developing Radioligand Binding Assays

There are certain assay parameters that need to be carefully evaluated for development of radioligand binding assays for natural product extract screening. These parameters are listed in Table **9**.

Table 9. Specific Assay Parameters to Consider for Developing Receptor Radioligand Binding Assays

- Receptor source and assay concentration
- Selection of radioligand and assay concentration
- Selection of assay incubation conditions
- Selection of assay buffer
- Selection of assay volume
- Selection of a methodology for separation of free and bound radioligand

2.4.2.1 Receptor source and assay concentration

2.4.2.1.1 Nonrecombinant Human Cell Lines

Stable non-recombinant human cell lines can often be used as a source of receptor for developing screening assays. Advantages of using this source include: cells contain unperturbed receptors, normal receptor post-translational modifications are usually present, and the physiologically activated signal transduction system is present for development of receptor functional assays. Possible disadvantages may include: the presence of undesired receptor subtypes, agonist-mediated receptor desensitization and internalization, and receptors may be structurally altered in transformed cells.

2.4.2.1.2 Membrane Preparations from Nonrecombinant Cells and Tissues

Crude plasma membranes prepared from human cells expressing receptors of interest are widely used for radioligand binding assays. Advantages of using membrane preparations include: agonist-mediated desensitization and internalization do not occur, large preparations can be made easily and frozen until needed, normal receptor postranslational modifications are present, and guanine nucleotides can be washed away to increase high affinity agonist binding to G-protein coupled receptors. Possible disadvantages include: undesirable receptor subtypes may be present and receptor-associated signal transduction molecules may not be present for developing functional assays.

2.4.2.1.3 Recombinant Receptor Containing Cells or Membranes

The primary advantage of using recombinant cells or membranes from recombinant cells as a source of receptors for radioligand binding assays is that they can provide a cheap, consistent, and pure population of human receptors for screening large numbers of compounds or extracts. Receptor densities expressed in cells can often be increased over physiological levels allowing increased ratios of signal-to-noise and the use of lower amounts of membranes. Large quantities of recombinant cells can be scaled up using spinner flasks or bioreactors, membranes can be prepared, and stored until needed for assays. Possible disadvantages of using recombinant cells or recombinant cell membranes include: cloning and expressing receptors of interest can be difficult and time consuming, over-expression of receptors may lead to activation of non-physiological signal transduction pathways, and physiological post-translational modifications may not occur in certain expression systems.

Several different expression systems can be used for recombinant human receptors including several types of mammalian cells, insect cells, yeast or bacteria. Important considerations for choosing a receptor expression system would include whether endogenous receptors are present, whether the appropriate receptor post-translational modifications will occur, whether appropriate signal transduction molecules are present for developing functional receptor assays, and which expression system will provide the largest quantity of receptor at the lowest cost.

2.4.2.1.4 Selection of Raioligand and Assay Concentration

For any given receptor, there are often several different radioligands that are commercially available for use in developing a radioligand binding assay. There are several important considerations that should be made when selecting a suitable radioligand and these considerations are listed in Table **10**.

Table 10. Considerations for Selecting a Radioligand

- Receptor affinity
- Specific activity
- Nonspecific binding
- Receptor selectivity
- Agonist or antagonist
- Type of radioisotope

The radioligand chosen should possess high affinity for the receptor so that the assay is sensitive, low concentrations of radioligand can be used to minimize assay costs, and levels of nonspecific binding can be minimized. Similarly, the chosen radioligand should have a high specific activity for similar reasons. High specific activity ligands are also essential when membrane preparations used contain low densities of receptor. The radioligand should also possess low nonspecific binding allowing maximization of signal-to-noise ratio. The chosen radioligand should also be selective for the subtype of receptor of interest, especially if related receptor subtypes are present in the receptor preparation used.

For many receptors, both radioactive agonists and antagonists may be commercially available for use in binding assays and in some cases, antagonists are preferred over agonists. For example, antagonists are generally preferred for developing binding assays for G-protein coupled receptors. The reason for this preference is due to the selective influence of endogenous guanine nucleotides on agonists binding. Only low affinity equilibrium agonist binding can be detected in whole cells due the presence of endogenous guanine nucleotides which promote conversion of high affinity to low affinity agonist binding sites (Thomsen *et al.*, 1988). In contrast, antagonist binding is not effected by guanine nucleotides and high affinity binding is detected. Therefore, antagonists are required for labeling binding studies using whole cells. Similarly, agonist binding to membranes containing significant levels of endogenous guanine nucleotides will also be reduced to low affinity. Due to this potential membrane contamination, during preparation antagonists are generally preferred for radioligand binding studies to G-protein coupled receptors. Antagonist radioligands would also be preferred for cytokine receptor binding assays because agonists promote rapid receptor internalization. However, no cytokine receptor antagonists are available for this application. Antagonist radioligands are also preferred for ligand-gated ion channel receptor binding assays because agonists can promote rapid receptor desensitization.

The last consideration in choosing a suitable radioligand is the choice of radioisotope. Tritiated or radioiodinated ligands are most commonly used in screening assays. It is important to note that radioiodinated ligands have a relatively short half-life and radioactive decay must be taken into account during use of the radioligand. Significant decay is often manifested in a binding assay by an increase in nonspecific binding. Table **11** lists some important characteristics of commonly used radioisotopes.

Table 10. Characteristics of Radioisotopes Used for Radioligand Binding Studies

Isotope Specific (Ci/mmol)	*Radioactivity*	*Half-Life (years)*	*Other Considerations*
[^{3}H]	40	12.3 years	Bioactivity of ligand usually unchanged with tritiation. Stable for long periods.
[^{125}I]	2125	60.2 days	Requires tyrosine or unsaturated cyclic system, especially used when there is low receptor density. No scintillation fluid is required. Generally only stable for a month. Biological activity can be altered with iodination.
[^{32}P]	9760	14.3 days	Short half life is a technical problem
[^{14}C]	0.06	5730 years	Specific activity too low

The final assay concentration of radioligand used in an assay is also an important variable in developing a radioligand binding assay. Typically, a K_d concentration of radioligand (1/2 of a saturating concentration) is used. It is possible that lower concentrations can be added if a high specific activity radioligand is available or if high receptor densities are present in the receptor preparation. Minimization of the radioligand concentration will reduce assay costs, minimize nonspecific binding, and reduce amounts of radioactive waste.

2.4.2.1.5 Selection of Assay Incubation Conditions

For radioligand saturation and competition experiments, the theoretical model used is one of equilibrium. Thus, the time of incubation needs to be sufficient to ensure that equilibrium or steady state radioligand binding is reached during the assay incubation (Bylund *et al.*, 1993). The period of time required to reach equilibrium depends upon radioligand affinity and concentration and the assay incubation temperature. For most radioligands with K_d values in the low nanomolar range, equilibrium is usually reached in 20-60 minutes at room temperature in the presence of a K_d concentration of radioligand. Most receptor binding assays can be conveniently performed at room temperature. However, there are exceptions. For example, cytokine binding assays are generally performed at 0-4 ^{o}C to reduce rapid agonist-mediated receptor internalization which occurs at higher temperatures. In addition, binding assays utilizing peptide radioligands are typically performed at room temperature or lower to minimize potential proteolysis that can occur at physiological temperatures. Inclusion of protease inhibitors is generally advisable when using peptide radioligands.

2.4.2.1.6 Selection of Assay Buffer

In general, radioligand binding assays are performed at physiological pH. There are several buffers commercially available that have good buffering capacity at physiological pH. Different buffers should be evaluated to determine which gives an optimal ratio of specific to nonspecific binding.

2.4.2.1.7 Selection of Assay Volume

The total assay volume for a radioligand binding assay should be minimized to conserve expensive reagents. Assay volumes ranging from 50 to 200 μl are typically used and volumes are generally limited by the precision of liquid handling devices used. In some cases where the receptor density is low, larger assay volumes up to 500-1000 μl may be required. New technologies are now being developed to enable dispensing of nanoliter quantities of reagents and it is likely that these technologies will be applied to reducing volumes of radioligand binding assays.

2.4.2.1.8 Selection of a Methodology for Separation of Free and Bound Radioligand

Subsequent to incubations, free and bound radioligand need to be resolved so that only bound ligand is measured. There are primarily three methods used, equilibrium dialysis, centrifugation, and vacuum filtration over glass fiber filters. For high volume screening, vacuum filtration is the only method of choice for resolving free and bound radioligand. The principle behind filtration is that filters will retain radioligand bound to membrane fragments or cells while unbound radioligand passes through the filter. Residual radioligand retained by filters is removed by subsequent washing of filters with ice cold buffer. One requirement for this washing procedure is that the dissociation of radioligand needs to be minimal during the washing step. Therefore, only radioligands with high affinity and slow dissociation rates can be used in filtration assays. Radioligands with affinities of less than 100 nM cannot generally be subjected to filtration assays because dissociation will occur in a fraction of a second (Yammamura *et al.*, 1985). Even after washing, considerable amounts of free radioligand can still be retained on filters resulting in increased levels of nonspecific binding. Most glass fiber filters are available with different pore sizes and it is advisable to test several sizes to determine which gives the lowest nonspecific filter binding. In certain cases, non-specific binding can also be reduced by chemical treatments of filters. For example, filter retention of positively charged radioligands to glass fiber filters can be reduced by precoating filters with polyethyleneimine (Bruns *et al*, 1983).

2.4.3 Steps in Developing and Validating Radioligand Binding Assays

The following section details the development and validation of a membrane radioligand binding assay for interleukin-8 receptor. This will serve as an example to illustrate many of the points concerning development of a radioligand binding assay made in the previous section. Interleukin-8 (IL-8) is an important pro-inflammatory chemokine that binds to human neutrophils and promotes neutrophil chemotaxis and degranulation (Kunkel *et al.*, 1991). The IL-8 receptor has been characterized pharmacologically and cloned and sequenced (Grob *et al.*, 1990; Thomas *et al.*, 1990; 1991; Moser *et al.*, 1991;Wu *et al.*, 1993). The IL-8 receptor is a member of the G-protein coupled receptor family that couples specifically with G-protein and stimulates phospholipaseβ (PLCβ) leading to enhanced membrane phosphatidylinositol hydrolysis and increases in intracellular concentrations of inositol-1,4,5-triphosphate

(IP_3), diacylglycerol (DAG), and intracellular calcium. Several pharmaceutical companies have developed IL-8 radioligand binding assays to screen for specific receptor antagonists that may be therapeutically useful as anti-inflammatory agents.

2.4.3.1 Cell Membrane Linearity Evaluation

The first step in developing a radioligand binding assay is to determine the optimal receptor concentration to include in the assay. This is accomplished by measuring total, specific, and nonspecific binding as a function of addition of increasing amounts of cells or membranes. Nonspecific binding is determined by measuring radioligand binding in the presence of a large excess concentration of unlabeled ligand (100-1000 fold higher than the radioligand K_d value). Specific binding is calculated by subtracting nonspecific binding from total binding. An optimal radioligand concentration is one where total binding is a relatively high signal (thousands of cpm) and the level of nonspecific binding is minimized.

Nonspecific binding can occur to filters as well as components of membranes or cells. After filter binding sites are saturated, nonspecific binding generally will increase linearly with increasing cell or membrane concentrations. Optimally, nonspecific binding should range between 0-20% for most radioligands. If nonspecific binding to filters is high, filter treatments designed to reduce this should be evaluated.

Human neutrophils were chosen as a source of human IL-8 receptors for developing the IL-8 radioligand binding assay. The decision was made to use crude membranes as a source of receptor instead of whole neutrophils so that large scale preparations of membranes could be made and stored until needed for screening. Neutrophils were isolated from human blood using Ficoll Paque and cells were disrupted using a liquid nitrogen cavitation bomb in the presence of a cocktail of protease inhibitors. Crude neutrophil plasma membranes were isolated by centrifugation at 40,000 X g. [^{125}I]IL-8 was used as the radioligand and no other alternatives were commercially available. A concentration of 250 nM IL-8 was selected to define nonspecific binding. Separation of bound and free [^{125}I]IL-8 was accomplished by vacuum filtration over PEI-treated Whatman GF/B glass fiber filters.

Results of a membrane-linearity experiment in which specific and nonspecific binding of 1 nM [^{125}I]IL-8 were measured in the presence of between 6 and 400 μg protein/tube are shown in Figure **32**. The ratio of specific binding to nonspecific binding was 9:1 at the highest membrane concentration tested. To conserve membranes, a concentration of 30 μg/tube was selected for subsequent assays. Nonspecific binding at this concentration was generally 20% of total binding giving an assay signal-to-noise ratio of approximately 5:1. The majority of nonspecific binding was determined to be to GF/B filters and was somewhat reduced by PEI treatment (data not shown).

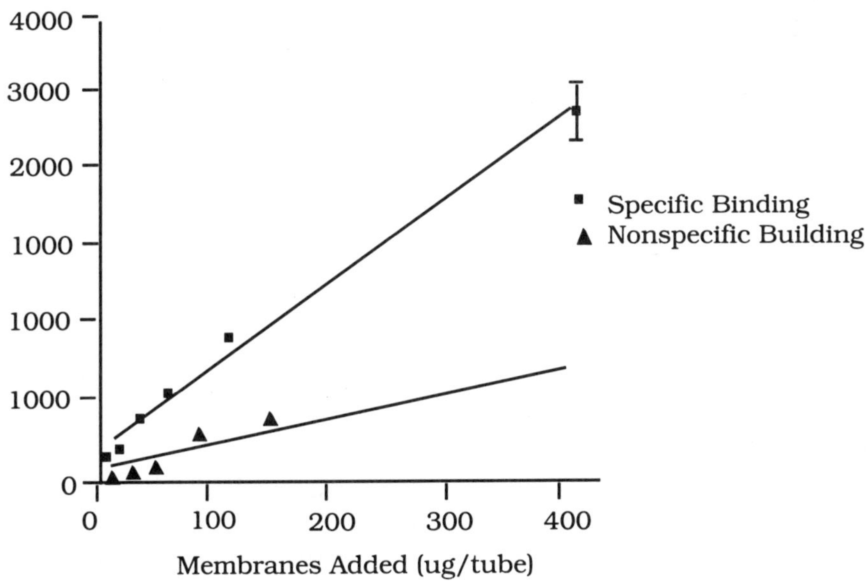

Figure 32: [^{125}I]IL-8 binding (1nM) to increasing concentrations of human neutrophil membranes. Incubations were conducted for 60 minutes at room temperature according to incubation conditions reported for [^{125}I]IL-8 binding to whole neutrophils (Grob *et al.*, 1990).

2.4.3.2 Association and Dissociation Time Courses

The primary reason for performing an association time course experiment is to determine when equilibrium radioligand binding is reached so that an incubation interval can be selected for the assay. Association time course experiments are simply performed by measuring the specific binding of a fixed concentration of radioligand as a function of the time of incubation. Figure **33** shows a representative plot of untransformed association time course data to illustrate the time when equilibrium is achieved.

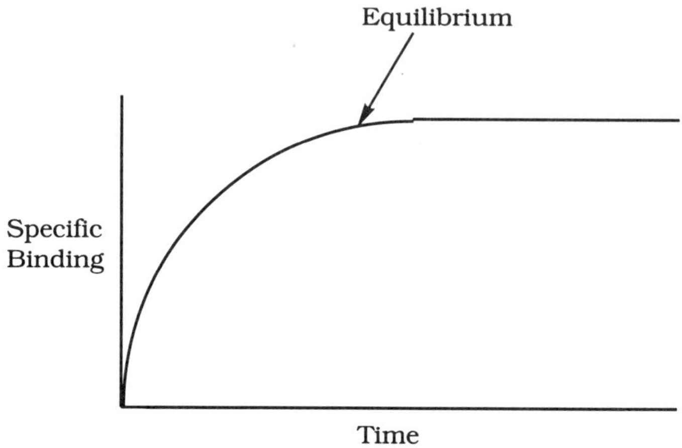

Figure 33: Representitive radioligand association time course data.

The association time course for a simple biomolecular reaction should consist of a single exponential component. The plateau of the time course indicates when equilibrium binding is achieved. At this point, the rate of radioligand association and dissociation are equal.

The above association time course binding data can be graphically transformed so that the observed rate constant (k_{obs}) can be determined. Figure **34** shows an example of such a transformation of association time course binding data where ln [*DR]eq/ ([*DR]eq-([*DR]) is plotted on the y-axis and time is plotted on the x-axis. k_1' is the pseudo first order

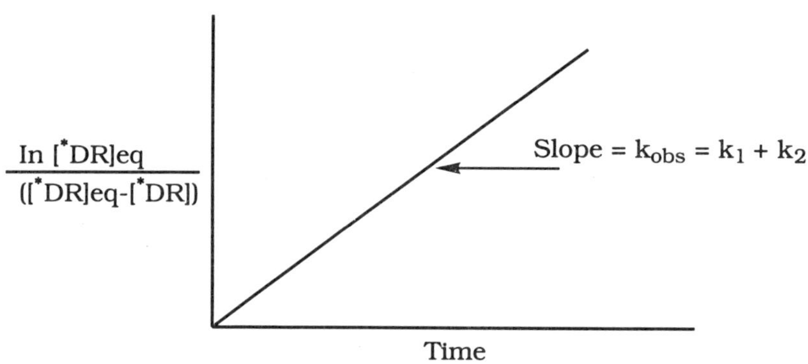

Figure 34: Transformation of representative association time course binding data.

association rate constant and k_2 corresponds to the dissociation constant. Deviations from a linear data transformation plot generally indicate that binding is not a simple biomolecular reaction and may indicate that cooperativity or multiple populations of binding site occur.

Data can be also be re-plotted to determine the association rate constant k_1 from k_{obs} if several association time courses are conducted at varying radioligand concentrations. The representative figure (Figure **35**) shown below represents how k_1 can be graphically determined.

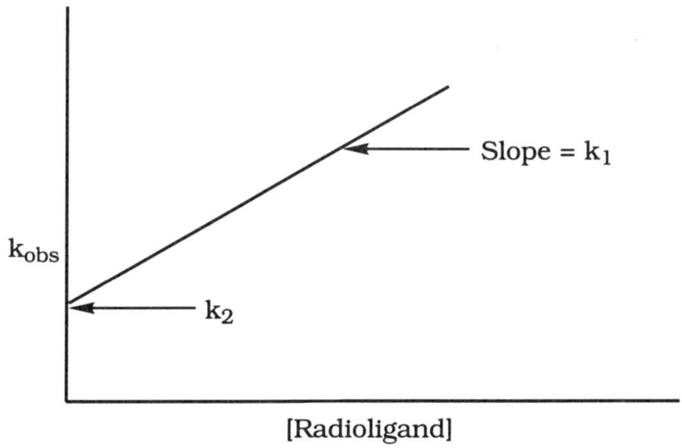

Figure 35: Replot of transformed association time course binding data.

A dissociation time course evaluation is also important to perform so that it can be established whether the ligand-receptor complex is sufficiently stable during filtration procedures. A dissociation time course experiment is performed by incubating cells or membranes and radioligand for a sufficient period to achieve equilibrium, addition of a 100-1000 fold excess on unlabeled ligand to initiate dissociation, followed by measurement of binding as a function of time. Dissociation can also be experimentally initiated by infinite dilution of the incubation mixture. Many investigators suggest performing dissociation time courses using both methods to initiate dissociation (Bylund *et al.*, 1993). Figure **36** shows a representative dissociation time course curve for a first order bimolecular species.

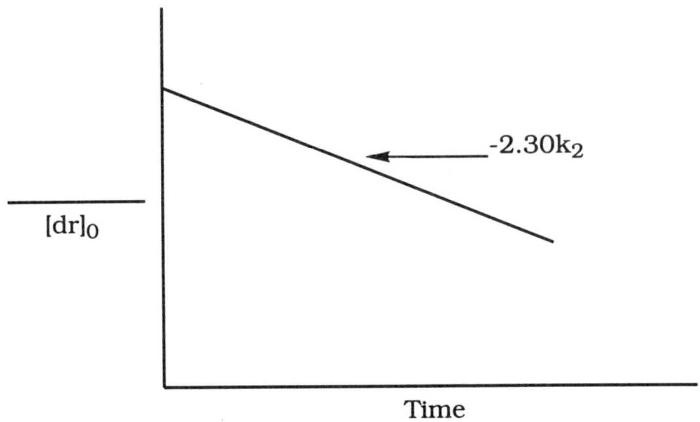

Figure 36: Transformation of dissociation time course binding data.

For radioligands that have half times of dissociation of 10 seconds or less, separation of free and bound ligand by filtration may not be appropriate due to instability of the ligand-receptor complex during washing steps.

Similar to the above general example, the association kinetics of specific $[^{125}I]$IL-8 binding to human neutrophil membranes was next evaluated to determine when equilibrium is reached and ultimately how long incubations should be conducted for assays (Figure **37**). A monoexponential association function using the nonlinear least squares curve fitting program called Prism (GraphPad, San Diego, CA). An estimate for the half-time of association 5 minutes was determined indicating relatively rapid association kinetics.

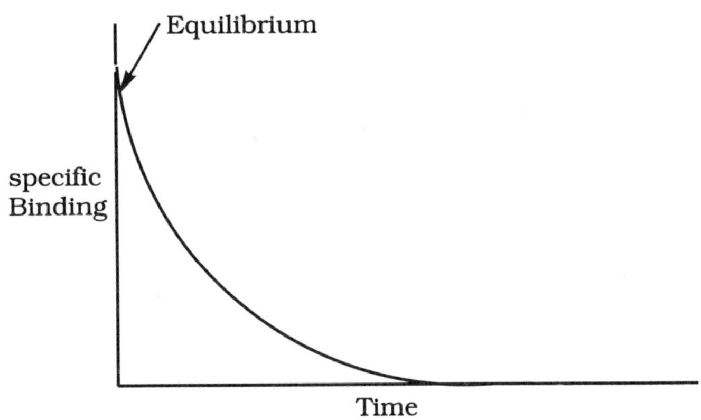

Figure 37: Representative dissociation time course binding data.

Transformation of dissociation time course data for a first order reaction by plotting. Log_{10} $[DR]/[DR]_o$ vs time will yield a linear plot with the slope being equal to $-2.303\ k_2$. Deviations from linearity may indicate positive cooperativity or the existence of multiple populations of receptors. A representation of this transformation is shown in Figure **38**.

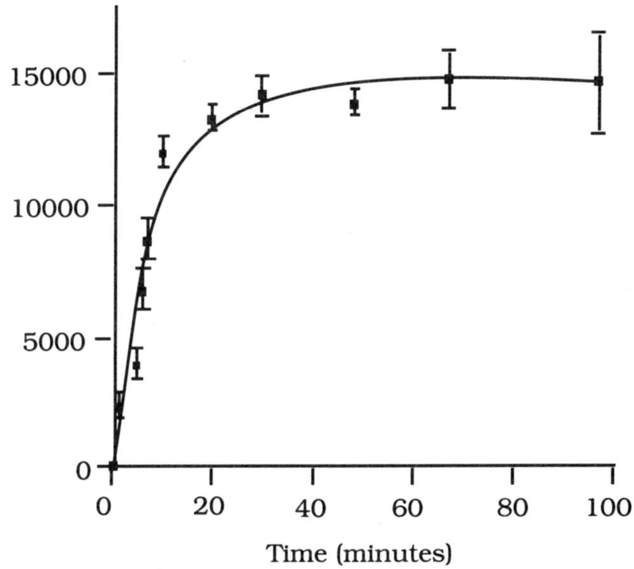

Figure 38: Association time course for [^{125}I]IL-8 binding to human neutrophil membranes. Concentrations of 1 nM [^{125}I]IL-8 and 30 ug/tube of neutrophil membranes were used in this experiment. Incubations were performed at room temperature for the indicated time periods. Data points represent the mean+/-SEM of triplicate determinations at each time point.

A dissociation time course experiment was also performed and dissociation data were found to best fit a monoexponential decay function using the Prism computer program (data not shown). An estimate of a half time for dissociation of 40 minutes was obtained indicating that negligible radioligand dissociation would be expected to take place during dilution with ice cold buffer and washing filters to remove unbound [^{125}I]IL-8 from filters.

2.4.3.3 Saturation Isotherm Evaluation

An important criterion for a valid radioligand binding assay is that binding must be saturable since a finite receptor density is predicted to be present. Saturation experiments are conducted by measuring total, specific, and nonspecific binding as a function of increasing radioligand concentration. Typically, radioligand concentrations tested should be in the range of 0.1 x K_d to 10 x K_d for the radioligand (Bylund *et al.*, 1993). When bound vs free radioligand is plotted for a simple bimolecular reaction, a hyperbolic curve is observed. The concentration of radioligand that gives maximal specific binding is defined as the K_d value and is used as a parameter of the affinity of a radioligand for receptor. A representative saturation isotherm is shown in Figure **38**.

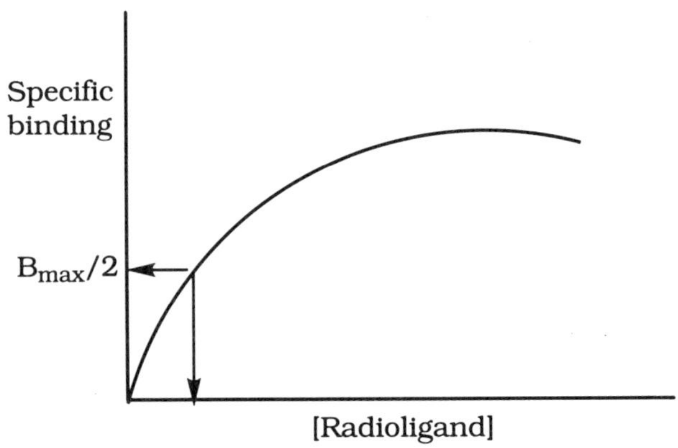

Figure 39: Representative radioligand saturation binding data for single population of receptors.

To determine experimental values for K_d and B_{max}, saturation binding data are often subjected to Scatchard analysis. In this case, values of bound/free vs bound radioligand are plotted on the y- and x-axes, respectively (Figure **39**). A linear plot is generally observed for radioligand binding to a single population of binding sites conforming to the simple law of mass action. The slope of the Scatchard plot provides an estimate of the K_d of the radioligand and the x-intercept provides an estimate of the receptor density or B_{max}. Curvlinear Scatchard plots may be indicative of either cooperativity or multiple populations of binding sites Figure **40**. For interprétation of curvelinear Scatchard plots and discussion of assumptions and possible artifacts associated with Scatchard transformation of saturation isotherm data, readers are encouraged to consult an excellent book written by Limbird, 1986. A more simple method to derive K_d and B_{max} values is to use a nonlinear regression analysis which is provided by a software package called Prism (GraphPad, San Diego, CA). Prism also allows fitting of data to one and two site binding equations to determine if the saturation data are best fit by a single or two populations of binding sites.

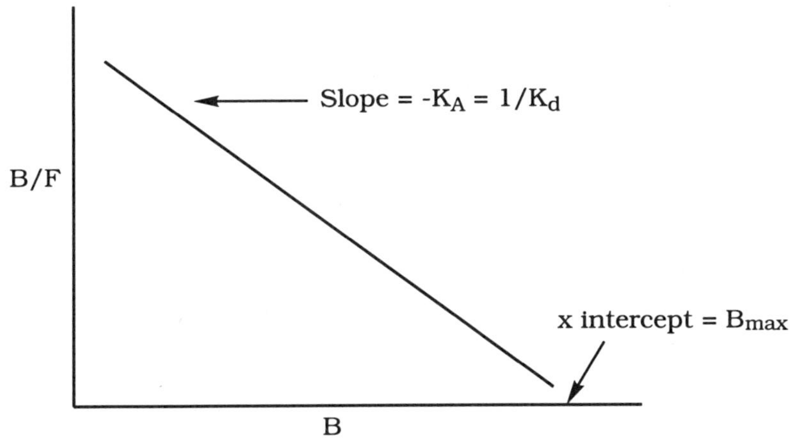

Figure 40: Scatchard transformation of saturation binding data, one population of binding sites.

Close evaluation of the experimental values of K_d and B_{max} are important in developing a radioligand binding assay. The experimentally determined K_d of radioligand should be close to expected value listed by the radioligand supplier and should be similar to the potency of the ligand derived from functional studies (unless spare receptors are present). The B_{max} value should also fall within expected limits for the tissue, membrane preparation, or cells studied. A B_{max} value higher than expected for cells or membranes could suggest that a portion of the radioligand binding measured is not due to the binding to the physiological receptor of interest.

Two independent saturation experiments for [^{125}I]IL-8 binding to human neutrophil membranes were performed during development of the assay and a representative saturation isotherm is shown in Figure **41**. Estimates for K_d and B_{max} values of 0.9+/-0.1 nM and 150-300 fmol/mg protein were determined using the Prism program. Data were also fit to a two-site model but the data were found to be best fit by a one-site model of binding. The K_d value obtained was in close agreement with values reported in the literature for whole cell binding studies but was an order of magnitude lower than the value reported from a study utilizing human polymorphonuclear leukocyte membranes (Barnett *et al.,* 1993).

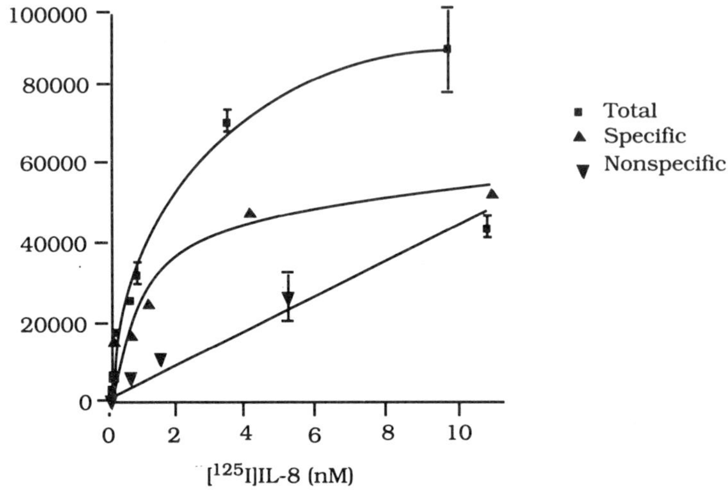

Figure 41: Saturation binding of [^{125}I]IL-8 to human neutrophil membranes. Assays contained 30 mg of membranes and incubations were performed for 60 minutes at room temperature. Data points represent the mean+/-SEM of triplicate determinations at each [^{125}I]IL-8 concentration.

2.4.3.4 Competition Studies

The specificity with which a radioligand interacts with a receptor is typically determined by competition studies. In these studies, membrane preparations or cells are incubated with a constant concentration of radioligand and increasing concentrations of unlabeled competitors. For any given receptor, detailed competition studies should be performed to demonstrate that the ligand selectivity expected for the receptor is exhibited for various relevant receptor agonists and antagonists. Generally, competition binding data are plotted as specific binding vs the $\log_{10}$ of the concentration of competitor. Again, the potency of a competing compound to inhibit radioligand binding should correlate with the ability to promote (if it is an agonist) or inhibit if it is an antagonist) the appearance of a physiological response associated with the receptor.

To quantitate the potency of a competing compound for radioligand binding sites, one determines the concentration of the competing compound that reduces radioligand binding by 50%. This concentration is referred to as the IC_{50}.The IC_{50} can be determined by nonlinear regression analysis using a suitable computer software package such as Prism (GraphPad, San Diego). After an IC_{50} value has been determined, the equilibrium dissociation constant for the competitor can be determined using the Cheng-Prussof equation:

$$K_{DI} = IC_{50} / (1 + [^{*}DR] / K_D)$$

where K_{DI} = equilibrium dissociation constant for competitor, I and K_D = equilibrium dissociation constant for radioligand

A major assumption for use of the Cheng Prussof equation is that [R] << K_D of the radioligand such that [*D] added = [*D]free. However, in certain cases the amount of added radioligand that binds to receptor may be greater than 5%. This could easily be the case in assays in which receptor concentration within the membrane is high and radioligand concentration added is of a similar magnitude. An alternative equation has been proposed by Linden (1982) for this situation in which greater than 5% added radioligand is bound:

$$K_1 = \frac{[I]}{1 + \frac{[^{*}D]}{K_D} + \frac{R_{TOT}}{K_D}\left[\frac{[K_D] + [^{*}D]/2}{K_D + [^{*}D]}\right]}$$

Where R_{TOT} = calculated concentration of binding sites

It is important to carefully evaluate the shape of competition curves during assay development and validation. For competition in which a single population of receptors is involved, the competition curve will proceed from 10% to 90% inhibition of radioligand binding over a 81-fold concentration of the competing compound. Curves that are more shallow and proceed from 10% to 90% over a greater than 81-fold range of competitor, have a more shallow shape which may be indicative of either negative cooperativity or interaction with multiple populations of binding sites. Conversely, a competing compound that reduces radioligand binding from 10% to 90% in less than a 81-fold concentration range, would have a steeper shape indicating possible positive cooperativity. One parameter commonly calculated to quantitate competition curve shape is the pseudo-Hill coefficient. Computer software packages such as Prism (GraphPad, San Diego, CA) can be used to calculate this coefficient. A normal steepness competition curve will have a pseudo-Hill coefficient of 1, a shallow curve will have a pseudo-Hill coefficient of less than 1, and a steep curve will have a pseudo-Hill coefficient greater than 1. It is important to stress that incubations during competition experiments must be sufficient not only for radioligand binding to achieve equilibrium but also sufficient for the competing compound to reach equilibrium. If this condition is not met, deviations from the normal steepness competition curve may occur (Ehlert *et al.*, 1981).

Several competition experiments were performed during development of the [^{125}I]IL-8 assay. Data from three experiments were pooled and the results are shown in Figure **42**.

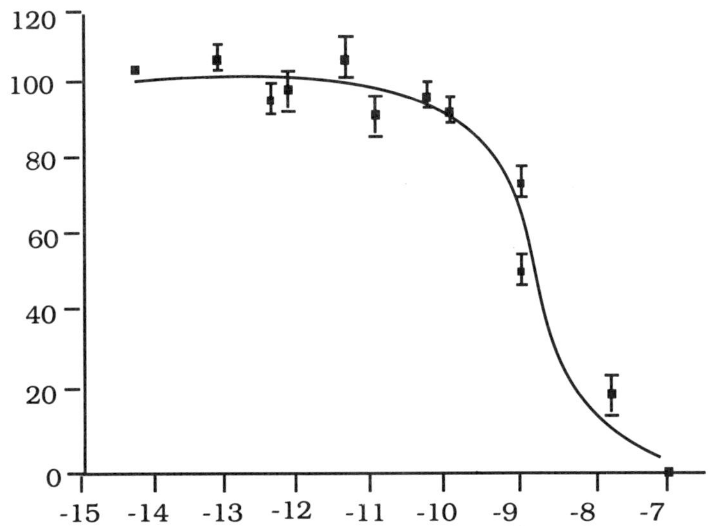

Figure 42: Competition of IL-8 for [^{125}I]IL-8 binding to human neutrophil membranes. 1 nM [^{125}I]IL-8 and 30 μg of membranes were present and incubated for 60 min at room temperature in the presence of indicated concentrations of IL-8. Data points represent the mean+/-SEM of data from three separate experiments in which triplicate determinations were made at each IL-8 concentration.

Competition data were fit best to a one site competition equation using the nonlinear least-squares program Prism. The reduction in radioligand binding occurred over approximately two log units as expected for a simple one-site model of binding. An IC_{50} value of 1.23+/-0.2 nM was determined which is in close agreement with saturation data. Several other cytokines were evaluated for their ability to compete for binding [^{125}I]IL-8 to neutrophil membranes. Saturating concentrations of IL-3, IL-5, IL-6, GM-CSF, FMLP, and C5a had no significant influence on [^{125}I]IL-8 binding as expected (data not shown).

Since the IL-8 receptor is a member of the seven transmembrane receptor superfamily, agonist binding should be reduced by addition of exogenous guanine nucleotides. Addition of the nonhydrolyzable GTP analog, GMP-PNP, was observed to reduce [^{125}I]IL-8 binding to human neutrophil membranes. However, only 60% of agonist binding was reduced suggesting that a subpopulation of the IL-8 receptor may not couple to G-protein or that membranes contain a larger population of receptors than G-protein (Figure **43**). Similar effects of guanine nucleotides on [^{125}I]IL-8 binding to human PMN membranes have been reported in the literature (Barnett *et al.*, 1993).

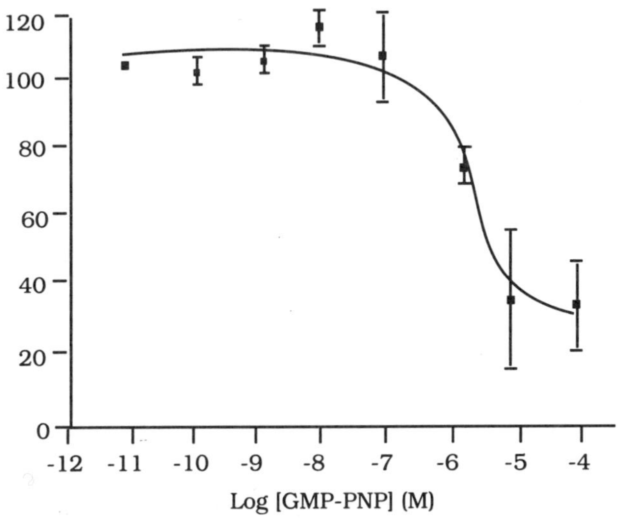

Figure 43: Effect of GMP-PNP on [^{125}I]IL-8 binding to human neutrophil membranes. Assays contained 1 nM [^{125}I]IL8, 30 µg of membranes and indicated concentrations of GMP-PNP and incubations were performed for 60 minutes at room temperature. Data represent the mean+/-SEM for triplicate determinations at each GMP-PNP concentration.

Similar to most radioligand binding assays, the IL-8 binding to human neutrophil membranes was not adversely affected by high concentrations of DMSO. A final assay concentration of up to 10% DMSO had no significant influence on [^{125}I]IL-8 binding (data not shown).

2.4.4 Alternative Radioligand Binding Assays

Although filtration radioligand binding assays have been widely used for drug discovery, new assay technologies have been commercially developed to simplify radioligand binding assays by eliminating the need to physically separate free and bound radioligand. These new higher throughput assays are being widely used by pharmaceutical and biotechnology companies and are largely replacing standard filtration radioligand binding assays in screening programs. Very little information is available concerning the suitability of these new assays to screening crude natural product extracts. Therefore, this suitability needs to be carefully assessed for different types of natural product extracts. The following section details some of these new assay methods and includes discussions of the basic principles, advantages and disadvantages of use in screening, how they are performed, and special instrumentation required for their use.

2.4.4.1 FlashPlate Assays

The FlashPlate™ technology, developed by Dupont-NEN, can be utilized to increase throughput of receptor radioligand binding assays in a screening program. The FlashPlate™ primarily consists of a 96-well microtiter plate which has been pre-coated with a specialized scintillant. The principle of the assay is that receptor becomes incorporated into the scintilant phase, radioligand binds to the receptor, and only bound radioligand associated with receptor incorporated into the scintillant will contribute to the assay signal. Membranes (with densities of 10 fmol receptor/well or greater) or purified soluble receptors can theoretically be used in the FlashPlate™ radioligand binding assay. Loading of purified receptors into the scintillant phase is much more efficient than loading of membranes. However, two potential problems may occur with the use of purified receptors. Association with polystyrene may inactivate the receptor or may introduce steric hinderance for the receptor binding site. These problems can sometimes be circumvented by precoating FlashPlates™ with non-neutralizing antisera to the receptor followed by loading of receptor. It is recommended that radioiodinated ligands be used initially and then tritiated radioligands can be tested and possibly substituted if sufficient signal-to-noise ratios are obtained. For some receptors, it may be possible to use the FlashPlate™ as a reaction vessel and centrifuge the FlashPlate™ at low speed to collect the receptor against the wall of the plate or let settle before counting. The counting efficiencies for different radioisotopes using the FlashPlate™ are lower than with traditional scintillation counting.

Radioligand binding assays utilizing the FlashPlate™ typically consist of the following assay steps (Figure **44**).

1) First, the receptor preparation is incorporated into the scintillant coated plates by centrifugation followed by washing to remove unincorporated receptor.
2) Assays are initiated by the addition of radioligand and test extracts followed by a sufficient incubation.
3) In some cases, the reaction mixture is removed by aspiration prior to counting. This would be advised in screening natural products extracts because removal of the extract will reduce potential extract quenching.
4) Plates are then be counted using either the Packard TopCount or the Wallac MicroBeta microtiter plate scintillation counters.

The primary benefits of using the FlashPlate™ for radioligand binding assays include: 96-well microtiter plate format amenable to automation, elimination of the filtration and scintillant addition steps, reduced volume of assays and radioactive waste, and more rapid counting in microtiter plate scintillation counters. The major drawback for use of FlashPlates™ is a reduced signal to noise-ratio compared to standard filtration binding assays utilizing standard scintillation counters. Because of this, the FlashPlate™ is best used for receptor preparations in which the receptor density is high. In addition, signal-to-noise is much better using high specific activity radioiodinated ligands instead of tritiated radioligands. Examples of the use of the FlashPlate™ for receptor radioligand binding assays have been reported for endothelin A and B receptors (Nichols *et al.*, 1992; Holland *et al.*, 1994) using radioiodinated peptide ligands.

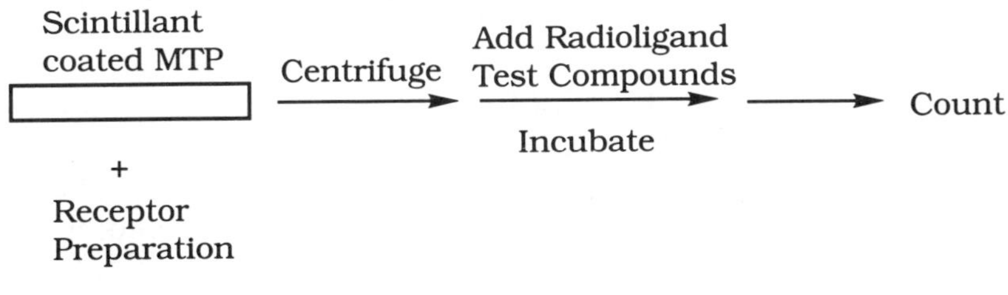

Figure 44: Steps associated with FlashPlate™ radioligand binding assays.

The major equipment required for high throughput screening using FlashPlates™ include:

-a microtiter plate centrifuge

-an automated liquid handling device (optional)

-a microtiter plate scintillation counter (Wallac MicroBeta or Packard TopCount)

-robotic arms for plate manipulation (optional)

2.4.4.2 Scintillation Proximity Assays

Amersham has developed a technology which has resulted in two simplification of receptor radioligand binding assays (Undenfriend *et al.*, 1987; Bosworth *et al.*, 1989). This technology is referred to as "Scintillation Proximity™" assays (SPA). The key component of the SPA assay are small fluoromicrospheres containing scintillant. Radioactive molecules brought into close proximity of the beads stimulate the scintillant to emit light which can be measured in a microtiter plate scintillation counter. Radioactive ligand that is too distant from the scintillant does not contribute to the signal. The majority of SPA assays utilize receptor membrane preparations that bind to wheat germ agglutinin (WGA) coated SPA beads.
The SPA assay consists of the following steps (Figure **45**):

1) [^{3}H]- or [^{125}I]-labeled ligand is added to microtiter plate wells. Assay incubations are initiated by addition of membranes.
2) WGA- or anti-receptor-specific antibody coated SPA beads are added followed by another incubation period.
3) Plates are then counted in a microtiter plate scintillation counter.

Advantages of the homogenous receptor SPA assay include; a 96-well microtiter plate format, omission of filtration and scintillant addition steps, and reduction in radioactive waste. Limitations of the SPA assay include use of only a [^{3}H]- or [^{125}I]-labeled ligands and lower signal-to-noise ratios than observed with standard radioligand filtration binding assays. Commercially available assay kits for angiotensin, endothelin, fibroblast growth factor, NPY, and IL6 receptors have been developed by Amersham and have been used in high throughput receptor screening programs using natural product extracts. All of these assay kits utilize radioiodinated ligands. The SPA technology also works for [^{3}H]-labeled ligands but requires membranes containing high densities of receptor. Quenching of the assay signal can occur with natural product extracts and can be corrected using a program supplied by Amersham. Besides selling specific kits, Amersham offers a technology transfer arrangement with pharmaceutical and biotechnology companies as well. Several papers describing the use of SPA technology for receptor binding assays have been published (Berry *et al.*, 1991; Hoffman *et al*, 1992; Jing *et al.*, 1992).

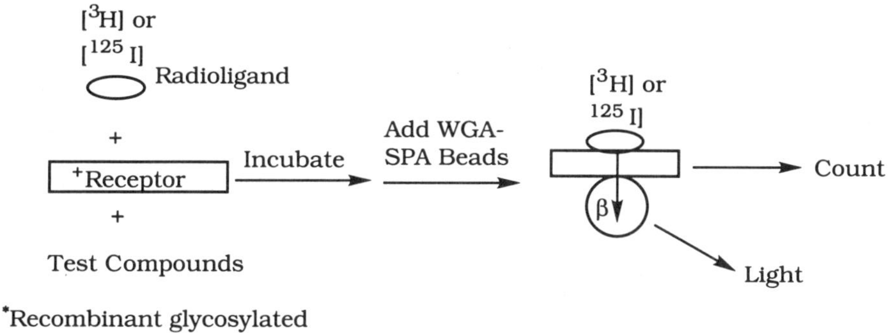

Figure 45: Steps involved in SPA receptor binding assays.

The major equipment required for high throughput screening using FlashPlates^{TM} include:

-an automated liquid handling device (optional)

-a microtiter plate scintillation counter (Wallac MicroBeta or Packard TopCount)

-robotic arms for plate manipulation (optional)

2.4.4.3 ScintiStrip Plate Assays

Wallac has developed a microtiter plate consumable in which scintillant is incorporated into the polystyrene plastic within the microtiter plate. These plates are called "ScintiStrip^{TM}" plates and can be coated with several materials including receptor antibodies, WGA, and steptavidin. Similar to FlashPlates^{TM}, only receptor bound radioligand associated with the scintillant contributes to the assay signal. Well-to-well cross-talk occurs with ScintiStrip^{TM} plates because they are transparent and Wallac has developed a cross-talk program which allows correction of the contribution of well-to-well cross-talk to the assay signal.

ScintiStrip™ plates offer the following advantages: they are arrayed in a 96-well microtiter plate format, no filtration and scintillant addition steps are required, radioactive waste is minimized, and several different radioiosotopes can be used. Receptor density may be an issue with use of these plates, especially with [^{3}H] ligands. Signal-to-noise is also reduced due to the correction of well-to-well cross talk.

The steps associated with the a representative ScintiStrip plate radioligand binding assay would be as follows:

1) Plates are coated with WGA, free WGA is washed away.
2) Receptor containing membranes are added and plates are incubated to allow receptor-WGA interaction to take place.
3) Radioligand and test extracts are added and plates are incubated to steady state.
4) Depending upon the radioligand, the reaction may be aspirated and plates are counted in a microplate scintillation counter.

Figure **46** details the steps that would be associated with a typical receptor binding assay utilizing membranes ScintiStrip™ plates. In this example, plates are precoated with WGA to promote interaction of glycosylated receptors in membranes with the plate well.

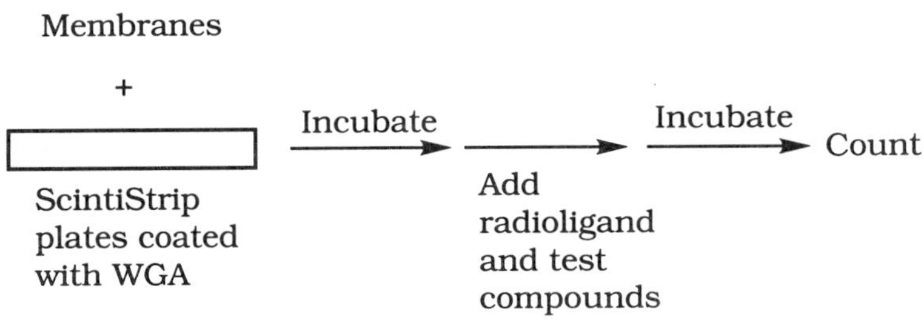

Figure 46: Assay steps associated with ScintiStrip™ receptor binding assays.

The major equipment required for use of ScintiStrip Plates™ for radioligand binding include:

-an automated liquid handling device (optional)

-a microtiter plate scintillation counter (Wallac MicroBeta only)

-robotic arms for plate manipulation (optional)

2.4.4.4 *Fluorescence Based Receptor Binding*

One important development concerning high throughput receptor assays is the increase in interest in fluorescence-based assays. Development of homogenous receptor binding assays utilizing fluorescent ligands will be an area of intense research and development by a number of different companies in the near future. Fluorescence detection can be a very sensitive: a compound can be excited up to 200,000 a second to produce nearly the same number of emitted photons. One potential problem with fluorescence detection is that background noise can be formidable. The progress made in this area is discussed in the following section.

2.4.4.4.1 Time Resolved Fluorescence

Fundamentally, time resolved fluorescence is based on the use of lanthanide chelate labels with unique fluorescence properties. These properties include a long fluorescence lifetime of the fluorescence signal which allows the measurement of the label fluorescence long after background fluorescence has declined. Lanthanide chelate labels also have a large difference between excitation and emission wavelengths which also results in large signal-to-noise ratios. Energy transfer enhanced lanthanide chelates such as chelates of europium, terbium, samarium, and dysprosium have this property and have been used for sometime in resolved fluorescence measurements (Scheinin *et al.*, 1994). Time resolved fluorescence has been widely used for immunoassays, detection of DNA probes and cytotoxicity assays. This technique has not widely been applied to measurement of ligand-receptor binding but one group has reported the measurement of ligand binding to membranes containing benzodiazepine receptors using a europium-labeled ligand as a tracer in a competitive binding assay. The bound ligand was measured from a centrifuged membrane pellet after resuspension into a special enhancement solution. Measurement of receptor-ligand complexes harvested by filtration, employing time resolved fluorescence, requires the use of very stable chelate labels and a specialized filter matrix with low luminescence background. It is anticipated that new technological developments will lead to further application of time resolved fluorescence to constructing sensitive fluorescence-based homogenous receptor binding assays.

2.4.4.4.2 Fluorescence Polarization

Fluorescence polarization is a relatively new homogenous assay method that can be used to evaluate protein-protein interactions such as binding of peptide ligands to peptide receptors. The fundamental features of this technology is that it can be used to evaluate the binding of a smaller molecule to a large molecule. It can also be used to determine the molecular weight of a molecule and has been used as a homogenous assay system for proteases. The basics of how fluorescence polarization works are as follows. Fluorescent molecules can be considered to have a definite axes whose directions can be compared to the directions of electrical vectors associated with absorption (excitation). Positive polarization results when a molecule absorbs and emits light in directions which may be considered parallel to one another. Under such conditions, excitation with vertically polarized light (polarized along the y-axis) produces fluorescence which has greatest intensity along the y-axis. The fluorescence polarization detector is used to take two readings of fluorescence intensity, one with the analyzer held vertically and another with it held horizontally. The polarization is defined as the difference of the two observed intensities over their sum. Polarization is also a function of rotation of molecules. The smaller a molecule, the faster it rotates and fluorescence polarization will be lower. Jolley Consulting and Research has developed a microtiter plate fluorescence polarimeter that can be used for micotiter plate based fluorescence polarization measurements and which is being used in high throughput drug screening programs. Several laboratories have used fluorescence polarization for measurement of ligand binding to receptors including binding of fluorescent-labeled EGF (Carraway *et al.*, 1993), fluorescent labeled glucagon (Heithier *et al.,* 1988; Tota *et al.*, 1995), fluorescent labeled neurokinins (Tota *et al.*, 1994;Turcatti *et al.*, 1995) to their perspective receptors. Important considerations for this application include a requirement for highly purified fluorescent tracer which is labeled with the fluorescent moiety in a proper position. Fluorescence polarization assays can be very high throughput and provide an excellent signal-to-noise ratio. Although this technology is relatively new, it is expected that it will be increasingly employed for receptor assays.

2.5 REFERENCES

Abe, K., Butcher, R. W., Nicholson, W. E., Baird, C. E., Liddle, R. A. and Liddle, G. W., *Endocrinology*, **84**, 362 (1969b).

Abe, K., Robison, G. A., Liddle, G. W., Butcher, R. W., Nicholson, W. E. and Baird, C. E., *Endocrinology*, **85**, 674 (1969a).

Akhtar, M. S., Athar, M. A. and Yaqub, M., *Planta Medica*, **42**, 205 (1981).

Alessi, D. R., Saito, Y., Campbell, D. G., Cohen, P., Sithanandum, G., Rapp, U. R., Ashworth, A., Marshal., C. J. and Cowley, S., *EMBO J.*, **13**, 1610 (1994).

Ali, L., Khan, A. K. A., Mamun, M. I. R., Moshiuzzaman, M., Nahar, N., Nur-e-Alam, M. and Rokeya, B., *Planta Med.*, **59**, 408 (1993).

Allen, L. F., Lefkowitz, R. J., Caron, M. G. and Cotecchia, S., *Proc. Natl. Acad. Sci.*, (USA), **88**, 11354 (1991).

Ames, B. N., Durston, W. E., Yamasaki E. and Lee, F. D., *Proc. Natl. Acad. Sci.,* (U.S.A.), **70,** 2281 (1973).

Aquino, R., De Feo, V., De Simone, F., Pizza, C. and Cirino, G., *J. Nat. Prod.*, **54**, 453 (1991).

Ascher, P. and Nowak, L., *J. Physiol.* **399**, 247 (1988).

Bank, U., Reinhold, D. and Ansorge, S. *Allerg. Immunol.*, **37**, 119 (1991).

Barnett, M.L., Lamb, K.A., Costello, K.M. and Pike, M.C., *Biochim. Biophys. Acta,* **1177**, 275 (1993).

Barrett, A. J., Kembhavi, A. A., Brown, M. A., Kirschke, H., Knight, C. G. and Hanada, M., *Biochem. J.*, **201**, 189 (1982).

Basha, G., Yap, P. and Penninckx, F., *Tumor Biology*, **17**, 354 (1996).

Bazzoni, F., Cassatella, M.A., Rossi, F., Ceska, M., Dewald, B., and Baggiolini, M., *J. Exp. Med.*, **173**, 771 (1991).

Beavo, J.A., *Physiol. Rev.*, **75**, 725 (1995).

Belser, W. L., Jr., Shaffer, S. D., Bliss, R. D., Hynds, P. M., Yamamoto, L., Pitts, J. N., Jr. and Winer, J. A., *Environ. Mutagen*, **3**, 123 (1981).

Benedek, G. and Szikszay, M., *Pharm. Res. Commun.*, **16**, 1009 (1948).

Beretz, A. and Cazenave, J-.P., "*Methods in Plant Biochemistry; Assays for Bioactivity*", (K. Hostettmann, ed.), Academic Press, London (1991), Volume **6**, pp. 235.

Bergammeyer, H. U., *Meth. Enz. Analysis*, Vol. **V**, 4 (1988).

Berridge, M.J., *Nature,* **361**, 315 (1993).

Berry, J.A., Burgess, A.J., and Towers, P., *J. Cardiovasc. Pharmacol.*, **17**, 143 (1991).

Blank, H. and Rewbell, G., *Arch. Derm.*, **92**, 319 (1965).

Bokoch, G.M., Katada, T., Northrup, J.K., Hewlett, E.L., and Gilman, A.G., *J. Biol. Chem.,* **258**, 2072 (1983).

Bosworth, N., and Towers, P., *Nature,* **341**, 167 (1989).

Brass, G., Shainhouse, J. Z. and Stevens, D. A., *Antimicrob. Agents. Chemother.*, **15**, 763 (1979).

Brauner-Osborne, H. B. and Brann, M. R., Eur. *J. Pharmacol.*, **295**, 93 (1995).

Bruns, R.F., Lawson-Wendling, K., Pugsley, T.A., *Anal. Biochem.*, **132**, 74 (1983).

Butcher, R.K., and Sutherland, E.W., *J. Biol. Chem.*, **237**, 1244 (1962).

Buttke, T. M., McCubrey, J. A. and Owen,T. C. *J. Immunol. Methods*, **157**, 233 (1993).

Bylund, D.B., Toews, M.L., *Am. J. Physiol.*, **265**, L-421 (1993).

Carpenter, C. L., Auger, K. R., Chaudhuri, M., Yoakim, M., Schaffhausen, B., Shoelson, S. F. and Cantley, L. C. *J. Biol. Chem.*, **268**, 9478.

Carraway, K.L., and Cerione, R.A., *Biochemistry,* **32**, 12039 (1993).

Casey, P.J., Solski, P.A., Der, C.J., and Buss, J.E., *Proc. Natl. Acad. Sci.* USA, **86**, 8323 (1989).

Cassel, D., and Selinger, Z., *Proc. Natl. Acad. Sci.* USA, **75**, 4155 (1979).

Castan'on, M.J., and Spevak, W., *Biochem. Biophys. Res. Commun.*, **198**, 626 (1994).

Chan, B. G., Waiss, A. C., Jr., Stanley, W. L. and Goodban, A. E., *J. Econ. Entomology*, **71**, 366 (1978).

Chapman, D. B. and Way, E. L., *Brit. J. Pharmacol.*, **75**, 389 (1982).

Chen, W.-J., Andres, D.A., Goldstein, J.L., Russell, D.W., and Brown, M.S., *Cell,* **66**, 327 (1991).

Cook, N., *Drug Disovery Today,* **1**, 287 (1996).

Cox, A.D., Hisaka, M.M., Buss, J.E., and Der, C.J., *Mol. Cell. Biol,.* **12**, 606 (1992).

Crabtree, G.R., and Clipstone, N.A., *Annu. Rev. Biochem.*, **63**, 1045 (1994).

Daniel, L. F., Thomas, R. B., Amal, M. B., David, T. C., Catherine, R. K., Donald, E. N., Daniel, F. O., Denis, J. S. and Jagadish, C. S., *J. Med. Chem.,* **34**, 518 (1991).

Daston, G. P., *Teratology*, **53**, 339 (1996).

DeLean, A., Stadel, J.M., and Lefkowtiz, R.J., *J. Biol. Chem.* **255**, 7108 (1980).

Delitheos, A. K., Papadimitriou, C. A. and Yannitsaro, A.G., *Fitoterapia*, **LXIII**, 441 (1992).

Denizot, F. and Lang, R. *J. Immunol. Methods*, **89,** 271 (1986).

Eberhand, M. Erne, P., *Biochem. Biophys. Res. Commun.*, **163**, 309 (1989).

Ebisawa, T., Karne, S., Lerner, M. R. and Reppert, S. M., *Proc. Natl. Acad. Sci.* USA, **91**, 6133 (1994).

Ehlert, F.J., *Mol. Pharmacol.*, **28**, 410 (1985).

Ehlert, F.J., Roeske, W.R., Yammamura, H.I., *Mol. Pharmacol.*, **19**, 367 (1981).

Einhelling, F. A., Leather, G. R. and Hobbs, L. L., *J. Chem. Ecol.*, **11**, 65 (1985).

Ellman, G. L., Courtney, D., Andres, J. and Featherstone, R. M., *Biochem. Pharmacol.,* **7**, 88 (1961).

Erne, P., Mittelhozer, E., Burgisser, E., Fluckiger, R. and Buhler, F. R. *J. Recept. Res.*, **4** 587, (1984).

Escobedo, J. A., Kaplan, D. R., Kavanaugh, W. M., Turck, C. W. and Williams, L. T., *Mol. Cell. Biol.*, **11**, 1125 (1991).

Ferrigni, N. R., Putnam, J. E., Anderson, B. L., Jacobsen, B., Nichols, D. E., Moore, D. S. and McLaughlin, J. L., *J. Nat. Prod.*, **45**, 679 (1982).

Fings, C. S., Tatliff, C. R. and Dunn, R. T., Glucose determination by *o*-toluidine method using acetic acid, "Clinical Chemistry" by Toro, C. and Ackerman, P. G., Little Browning and Company, Boston, 115 (1970).

Fournet, A., Barrios, A. A. and Munoz, V., *J. Ethnopharmacol.*, **41 (1-2)**, 19 (1994).

Fricker, S. P. and Buckley R. G., *Anticancer Reserch*, **16**, 3755.

Geoffroy, F. J., Allegra, C. J., Sinha, B. and Grem, J. L., *Oncology Research*, **6** (12), 581-591 (1994).

Gerlier, D. and Thomasset, N. *J. Immunol. Methods*, **94**, 57 (1986).

Gibbs, J. B. and Oliff, A., *Cell.*, **79**, 193 (1994).

Gibbs, J. B., Schaber, M. D., Allard, W. J., Sigal, I. and Scolnick, E. M., *Proc. Natl. Acad. Sci.*, (USA), **85**, 5026 (1988).

Gil, D. W. and Wolfe, B. B., *J. Pharmacol. Exper. Ther.*, **232**, 608 (1985).

Gressner, A. M., Polzar, B., Lahma, B. and Mannherz, H. G. *Hepatology* **23**, 571 (1996).

Grob, P. M., David, E., Warren, T. C., DeLeon, R. P., Farina, P. R., and Homon, C. A. *J. Biol. Chem.*, **265**, 8311 (1990).

Gupta, R., Dhanjal, R. and Madhubala, R., *Journal of Parasitic Diseases*, **19**, 49 (1995).

Hamburger, M. O. and Cordell, G. A., *J. Nat. Prod.*, **50**, 19 (1987).

Hanks, J. H. and Wallace, R. E., *Proc. Soc. Exp. Biol. Med.*, **71**, 196 (1949).

Hearing, V. J., "*Methods in Enzymology*", **142**, pp. 154-165, Academic Press, New York (1987).

Heeg, K., Reimann, J., Kabelitz, D., Hardt, C. Wagner, H., *J. Immunol. Methods*, **77**, 237 (1985).

Heithier, H., Ward, L.D., Cantrill, R.C., Klein, H.W., Im, M.J., Pollak, G., Freeman, B., Schiltz, E., Peters, R., Helmreich, E.J.M., *Biochim. Biophys. Acta*, **971**, 298 (1988).

Heldin, C. H., *Trends Biochem. Sci.*, **16**, 450 (1991).

Hemmila, I., and Harju, R., "Bioanalytical Applications of Labeling Technologies". Eds. Hemmia, I., Stahlberg, T. and Mottram, P., Wallac Oy, Turku Finland, p 84-119, 1994.

Hesketh, T. R., Smith, G. A., Moore, J. P., Taylor, M. V. and Metcalfe, J. C. *J. Biol. Chem.*, **258**, 4876 (1983).

Hoffman, R. and Cammeron, L., *Anal. Biochem.,* **203**, 70, (1992).

Holland, J.D., Singh, P., Brennand, J.C., and Garman, A.J., *Anal. Biochem.*, **222**, 516 (1994).

Homans, A. L. and Fuchs, A., *J. Chromatog.*, **51**, 327 (1970).

Hostettmann, K., Kizu, H. and Tomimori, T., *Planta Medica*, **44**, 34 (1982).

Hoth, M., and Penner, R., *Nature,* **355**, 353 (1992).

Isman, M. B., "*Insecticides of Plant Origin*" (Arnaen, J. T., Philogene, B. J. R. and Morand, P., Eds.), ACS Symposium Series 387, American Chemical Society, Washington, DC (1987), pp. 44.

Isman, M. B., Proksch, P. and Yan, J.-Y., *Entomol. Exp. Appl.*, **43**, Dr. W. Junk Publishers, Dordrecht (1987), pp. 87-93.

Jacobs, R. S., White, S. and Wilson, L., *Fed. Proc.*, **39**, 26 (1980).

Jakobs, K.H., Aktories, K., and Schultz, G., *Adv. Cyclic Nucl. Res.*, **14**, 173 (1981).

Jhon, D.-Y., Lee, H.-H., Park, D., Lee, C.-W., Lee, K.-H., Yoo, O.J., Rhee, S.G., *J. Biol. Chem.*, **268**, 6654 (1993).

Jing, S., Tapley, P., and Barbacid, M., *Neuron,* **9**, 1067 (1992).

Julien, J., Gasquet, M., Maillard, C., Balansard, G. and Timon-David, P., *Planta Medica*, **51**, (**3**), 205-208 (1985).

Julien, J., Gasquet, M., Maillard, C., Balansard, G. and Timon-David, P., *Planta Medica*, **51**, 205 (1985).

Julius, D., Livelli, Y. J., Jessel, T. M. and Axcl, R., *Science*, **244**, 1057 (1989).

Kakegawa, H., Matsumoto, H., Endo, K., Satoh, T., Nonaka, G. and Nishioka, I., *Chem. Pharm. Bull.,* **33**, 5079 (1985).

Kao, J. P., Harootunian, A. T., Tsien, R. Y., *J. Biol. Chem.*, **264**, 8179 (1989).

Karmen, A., Wroblewskis, F. and Ladue, J., *J. Clin. Invest.*, **34**, 126 (1955).

Kavanagh, F., "Analytical Microbiology" (Kavanagh, F., ed.), Academic Press, London (1963), pp. 125-141.

Kawai, T., Kinushita, K., Koyama, K. and Takahashi, K., *Planta Medica*, **60**, 17 (1994).

Kawashima, K., Miwa, Y., Kimura, M., Mizutani, K., Hayashi, A. and Tanaka, O., *Planta Medica*, **3**, 187 (1985).

Kawazu, K. *Advances in Natural Products Chemistry* (Natori, S., Itekawa, N. and Suzuki, M., Eds.), John Wiley, New York (1981), pp. 249.

Kazmi, S. U., Siddiqui, R. and Shekhani, S. *Frontiers in Natural Products Chemistry* (Atta-ur-Rahman, Ed.), Shamim Printing Press, Karachi (1990), p. 739-754.

Keene, A. T., Harris, A., Phillipson, J. D. and Warhurst, D. C., *Planta Medica*, **52**, 278 (1986).

Kiso, Y., Tohkin, M. and Hikino, H., *J. Nat. Prod.,* **46**, 841 (1983).

Kiso, Y., Tohkin, M. and Hikino, H., *Planta Medica*, **49**, 222 (1983).

Koch, C. A., Anderson, D., Moran, M. F., Ellis, C. and Pawson, T., *Science*, **252**, 668 (1991).

Kolb, A.J., and Neumann, K., *J. Bimolec. Screening,* **1**, 85 (1996).

Kolch, W., Heidecker, G., Kochs, G., Hummel, R., Vahidi, H., Mischak, H., Finkenzeller, G., Marme, D. and Rapp, U. R., *Nature*, **364**, 249 (1993).

Kong, Y. C., Lau, C. P., Wat, K. H., But, P. P. H., K. Cheng, F. and Waterman, P. G., *Planta Medica*, **55**, 176 (1989).

Kong, Y. C., Wat, K. H., Wong, A., Sexena, I. F., Cheng, K. F., But, P. P. H. and Chang, H. T., *Planta Medica*, **51**, 304 (1985).

Kubo, I., "*Method in Plant Biochemistry*" (Dey, P. M. and Harborne, J. B., Eds), Hostettman, K. Academic Press, London, **6**, 179 (1991).

Kubo, I., "*Methods in Plant Biochemistry*" (Dey, P. M. and Harborne, J. B., Eds), Hostettman, Academic Press, London, **6**, (1991) pp. 185.

Kunkel, S., Standiford, T., Chensue, s. W., Kasahara, K. and Strieter, R. M., *Agents Actons Suppl.*, **32**, 205 (1991).

Lee, C.-W., Lee, K.-H., Lee, S.B., Park, D., and Rhee, S.G., *J. Biol. Chem.*, **269**, 25335 (1994).

Lee, C.H., Park, D., Wu, D., Rhee, S.G., and Simon, M.I., *J. Biol. Chem.*, **267**, 16044 (1992).

Lee, S.B., and Rhee, S.G., *Curr. Opin. Cell Biol.* **7**, 183 (1994).

Lerner, M. R., *Trends in Neurosci.*, **17**, 142 (1994).

Leven, M., Vanden B., D. A., Mertens, F., Vlietinck, A. and Lammens, E., *Planta Medica*, **36**, 311, (1979).

Limbird, L.E., "Cell Surface Receptors: A Short Course on Theory and Methods". Boston, MA: Martinus Nijhoff, 1986.

Linden, J., *J. Cyclic Nucleotide Res.,* **8**, 163 (1982).

Linton, A. H., "Antibiotics: Assessment of Antimicrobial Activity and Resistance" (Russell, A. D. and Quernel, L. B., Eds), Academic Press, London (1983), pp. 19-30.

Long-term Effects of Pilocarpine in Rats: *Structrual Damage of the Brain Triggers Kindling and Spontaneous Recurrent Seizures.* Cavalheiro, E. A., Leite, J. P., Bortolotto, Z. A., Turski, W. A., Ikonomidou, C. and Turkshi, L., *Epilepsia*, **32**, 778-783.

Lowry, D. R. and Willumsen, *Annu. Rev. Biochem.*, **62**, 851 (1993).

Malgaroli, A., Milani, D., Meldolesi, J. and Pozzan, T., *J. Cell. Biol.*, **105**, 2145 (1987).

Mamber, S. W., Okasinski, W. G., Pinter, C. D. and Tunac, J. B., *Mutation Research*, **171**, 83, (1986).

Margolis, B., *Cell Growth Differ*, **3**, 73 (1992).

Maron, M. D. and Ames, B. N., *Mutation Research.* **113**, 173, (1983).

Mayer, B. J. and Baltimore, D., *Trends Cell Biol.*, **3**, 8 (1993).

McClintock, T. S., Graminski, G. F., Potenza, M. N., Jayawickreme, C. K., Roby-Shemkovitz, A. and Lerner, M. R., *Anal. Biochem.*, **209**, 298 (1993).

Merritt, J. E., McCarthy, S. A., Davies, M. P., Moores, K. E., *Biochem. J.*, **269**, 513 (1990).

Messier, T. L., Dorman, C. M., Brauner-Osborne, Eubanks, D. and Brann, M. R., *Pharmacol. Toxicol.*, **76**, 308 (1995).

Meyer, B. N., Ferrigini, N. R., Putnam, J. F., Jacobson, L. B. Nichols, D. E. and McLaughlin, J. L., *Planta Medica*, **45**, 31 (1982).

Mgonyebi, O. P., Russo, J. and Russo, I. H., *Int. J. Oncol.*, **12**, 865 (1998).

Migeon, J.C., and Nathanson, N.M., *J. Biol. Chem.*, **269**, 9767 (1994).

Minta, A., Kao, J. P., Tsien, R. Y., *J. Biol. Chem.*, **264**, 8171 (1989).

Moser, B., Schumacher, C., von Tscharner, V., Clark-Lewis, I., Baggiolini, M., *J. Biol. Chem.*, **266**, 10666 (1991).

Murray, M.V., Lalan, P., Gill, M.F., Addo, J.M., Lewis, D.K.H., Lee, R., Rampulla, M.P., Wachter, J.D., His, J.D., and Underwood, D.C., *Bioorg. Med. Chem. Lett.*, **2**, 1775 (1992).

Musacchio, A., Gibson, T., Lehto, v. -P. and Saraste, M., *FERS Lett.*, **307**, 55 (1992).

Neubig, R.R., Gantzos, R.P., and Braiser, R.S., *Mol. Pharmacol.*, **28**, 475 (1985).

Neubig, R.R., Gantzos, R.P., and Thomsen, W.J., *Biochemistry,* **27**, 2374 (1988).

Nichols, J.S., LeVine, H., Smith, G.F.H., Wypij, D.M., and Wiseman, J.S., *Biochem. Biophys. Meth.*, **25**, 173 (1992).

Nishizuka, Y., *Science* **258**, 607 (1992).

O'Callaghan, C. H., Morris, A., Kirby, S. M. and Shingler, A. H., *Antimicrob. Agents Chemother*.,**1**, 283 (1972).

Overduin, M., Rios, C. B., Mayer, B. J., Raltimore, D. and Cowburn, D., *Cell.*, **70**, 679 (1992).

Pawson, T., *Curr. Opin. Genet. Dev.*, **2**, 4 (1992).

Perros, P. and Weightmann, D. R., *Cell Prolif.*, **24**, 517 (1991).

Pomerantz, S. H., *J. Biol. Chem.,* **238,** 2351 (1963).

Pompliano, D. L., Rands, E., Schaber, M. D., Mosser, S. D., Anthony, N. J. and Gibbs, J. B., *Biochemistry*, **31**, 3800 (1992).

Potenza, M. N., Graminski, G. F. and Lerner, M. R., *J. Biol. Chem.,* **268** (8), 5959 (1993).

Potenza, M. N., Graminski, G. F., Schmauss, C., Lerner, M. R., *J. Neurosci.*, **14** (3), 1463, (1994).

Potenza, M. N., Lerner, M. R., *Pign. Cell Res.*, **4** (4), 186, (1991).

Putney, J.W.J., *Cell Calcium,* **11**, 611 (1990).

Rana, A. C. and Avadhoot, Y., *Fitoterapia,* **LXIII**, 60 (1992).

Reiss, Y., Goldstein, J.L., Seabra, M.C., Casey, P.J., and Brown, M.S., *Cell,* **62**, 81 (1990).

Reissig, J. L., Strominger, J. L. and Leloir, L. F., *J. Biol. Chem.*, **217**, 957 (1955).

Roehm, N. W., Rogers, G. H., Hatfield, S. M. and Glasebrook, A. L., *J. Immunol. Methods*, **142**, 257 (1991).

Rozdzial, M. M. and Ilaimo, L. T., *Cell*, **47**, 1061 (1986).

Sassone-Corsi, P., Visvader, J., Ferland, L., Mellon, P.L., Verma, I., *Genes Dev.*, **2**, 1529 (1988).

Scanlon, M., Williams, D. A. and Fay, F. S. *J. Biol. Chem.*, **262**, 6308 (1987).

Schafer, W. R. and Rine, J., *Annu, Rev. Genet.*, **26**, 209 (1992).

Schales O. and Schales, S. S., *J. Biol. Chem.*, **140**, 879, (1941).

Schlessinger, J., *Trends Biochem. Sci.*, **13**, 443 (1988).

Schroeder, K.S., and Neagle, B.D., *J. Biomolec. Screening,* **1**, 75 (1996).

Schwarts, O. S., Henin, Y., Marechal, V. and Montagnier, L., *AIDS Res. Hum. Retrovir.*, **4**, 441 (1988).

Seglen, P. O., *Methods Cell Biol.*, **13**, 29 (1976).

Sheinin, M., Sjoholm, B., Webb, S., Kovanen, S., and Hemmila, I., "Bioanalytical Applications of Labeling Technologies". Eds. I. Hemmia, T. Stahlberg, and P. Mottram. Wallac Oy, Turku Finland, p 250-261, 1994.

Shigeharu, N., Naoki, A. and Tadashi, S., *Biol. Pharm. Bull.*, **18** , 1145 (1995).

Snedecor, G. W., *Statistical Methods*, 5th edition, The Iowa State University Press, Ames, Iowa, USA (1965).

Solis, P. N., Wright, C. W., Anderson, M. M., Gupta, M. P. and Phillipson, J. D., *Planta Medica*, **59**, 250 (1993).

Spontaneous Recurrent Seizures in Rat: An Experimental Model of Partial Epilepsy, Leitz, J. P., Bortolotto, Z. A. and Cavalheiro, E. A., *Neuroscience & Biobehavioral Reviews*, **14**, 511-57 (1990).

Squire, B. J. and Whitfield, P. J., *Phytotherapy Research*, **3**, 112 (1989).

Srivastava, O. P., "Techniques for the Evaluation of Antimicrobial Properties of Natural Products" in *The Use of Pharmacological Techniques for the Evaluation of Natural Products* (Dhawan, B. N. and Srimal, R. C.) UNESCO, p. 72 (1984).

Stein, U., Walther, W. and Shoemaker, R. H., *J. Nat. Can. Inst.*, **88**, (19), 1383-1392 (1996).

Stuehr, D. J. and Marletta, M. A., *Proc. Natl. Acad. Sci.*, (USA) **82**, 7738 (1985).

Stuehr, D. J., Cho, H. J., Kwon, N. S., Weise, M F. and Nathan, C. F., *Proc. Natl. Acad. Sci.*, (USA), **88**, 7773 (1991).

Sugden, D. and Rowe, S. J., *J. Cell Biol.*, **119**, 1515 (1992).

Tachibana, Y. and Kawanishi, K., *Planta Medica*, **58**, 250, (1992).

Tajima, T., Hironao, T., Kajikawa, T. and Kawamura, H. *J. Vet. Med. Sci.*, **54**, 1187 (1992).

Tanaka, S., Yoon, Y., Fukui, H., Tabata, M., Akira, T., Okano, K., Iwai, M., Iga, Y. and Yokoyama, K., *Planta Medica*, **55**, 245,(1989).

Tang, W. -J., Iniguez-Lluhi, J. A., Mumby, S., and Gilman, A. G., *Cold Spring Harbor Symp. Quant. Biol.*, **57**, 135 (1992).

Tanred, S., Hartmut, K. and Samud, M. R., "Prostaglandins and Related Substances", IRL Press Limited, Oxford, England, 229 (1987).

Taussig, R., and Gilman, A. G., *J. Biol. Chem.*, **270**, 1 (1995).

Taylor, A. L. and Sasser, J. N. *Biology, Identification and Control of Root-Knot Nematode, Meloidogyne Spp.,* North Carolina State University, Graphics Raleigh, N. C., USA (1978).

Taylor, S. J., Chae, H. Z., Rhee, S. G., and Exton, J. H., *Nature,* **350**, 516 (1991).

Tecle, II., Lauffer, D. J., Davis, R. E., Miradegan, T., Moreland, D. W., Schwarz, R. D., Thomas, A. J., Raby, C., Eubanks, D., Brann,

M. R. and Jacn, J. C., *Bioorg. Medicinal Chem. Lett.*, **5**, 637 (1995).

Thomas, K. M., Pyun, H. Y., and Navarro, N. J., *J. Biol. Chem.*, **265**, 20061 (1990).

Thomas, K. M., Taylor, L., and Navarro, N. J., *J. Biol. Chem.*, **266**, 14839 (1991).

Thomsen, W. J., Jaquez, J., and Neubig, R. R., *Mol. Pharmacol.*, **34**, 814 (1988).

Tota, M. R., Daniel, S., Sirotina, A., Mazina, K. E., Fong, T.M., Longmore, J., and Strader, C.D., *Biochemistry,* **33**, 13079 (1994).

Tota, M. R., Xu, L., Sirotina, A., Strader, C. D., Graziano, M.P., *J. Biol. Chem.*, **270**, 26466 (1995).

Triesmann, R., Curr. Opin. Genet. Dev. 4: 96-101 (1994). 22. Schindler, C. and Darnell, J. E., *Annu. Rev. Biochem.*, **64**, 621 (1995).

Tsien, R. Y., Pozzan, T. and Rink, T. J., *J. Cell. Biol.*, **94**, 325 (1982).

Turcatti, G., Vogel, H., and Chollet, A., *Biochemistry,* **34**: 3972 (1995).

Ueno, II., Colbert, H., Escobedo, J. A. and Williams, L. T., *Science*, **252**, 844 (1991).

Undenfriend, S., Gerber, L., and Nelson, N., *Anal. Biochem.*, **161**, 494 (1987).

Van der Geer, P., Hunter, T. and Lindberg, R. A., *Annu. Rev. Cell. Biol.*, **10**, 251 (1994).

Van der Geer, P., Wiley, S., lai, V. K., Olivier, J. P., Gish, G. D., Stephens, R., Kaplan, D., Shoelson, S. and Pawson, T., *Curr. Biol.*, **5**, 404 (1995).

Van Weerden, Moerings, E. P., van Kreuningen, A., de Jong, F. H., van Steenbrugge, G. J. and Schroder, F. H., *Cell Prolif.*, **26**, 67 (1993).

Villar, A., Paya, M., Hortigüela, M. D. and Cortes, D., *Planta Medica*, **52,** 43 (1986).

Wagner, S., Beil, W., Westermann, J., Logan, R. P., Bock, C. t., trautein, C., Bleck, J. S. and Manns, M. P., *Gastroenterology*, **113**, 1836 (1997).

Weichert, H., Blechschmidt, I., Schroder, S. and Ambosius, H. *Allerg. Immunol.*, **37**, 139 (1991).

Weislow, O. W., Kiser, R., Fine, D., Bader, J., Shoemaker, R. H. and Boyd, M. R., *J. Natl. Cancer Inst.*, **81**, 577 (1989).

White, S. J. and Jacobs, R. S., *Mol. Pharmacol.,* **20**, 614-620 (1981).

Williams, N. G., Roberts, T. M. and Li, P., *Proc. Natl. Acad. Sci.*, (USA), **89**, 2922 (1992).

Wreggett, K.A., and DeLean, A., *Mol. Pharmacol.*, **26**, 214 (1984).

Wu, D., LaRosa, G. J., and Simon, M. I., *Science,* **261**, 101 (1993).

Yagi, A., Kanabara, T. and Morinabu, N., *Planta Medica*, **53**, 515 (1987).

Yamahara, J., Mochizuki, M., Rong, H. Q., Matsuda, H. and Fujimura, H., *J. Ethnopharm.*, **23**, 299 (1988).

Yammamura, H. I., Enna, S. J., and Kuhar, M. J. (Eds.). Neurotransmitter Receptor Binding (2nd ed.). N. Y. Raven, 1985.

Zarroug, M. A., Nugud, A. D., Bashir, A. K. and Mageed, A. A., *Int. J. Crude Drug Res.,* **26**, 77 (1988).

Subject Index.